T0155499

Signals and Communication Technology

The series "Signals and Communications Technology" is devoted to fundamentals and applications of modern methods of signal processing and cutting-edge communication technologies. The main topics are information and signal theory, acoustical signal processing, image processing and multimedia systems, mobile and wireless communications, and computer and communication networks. Volumes in the series address researchers in academia and industrial R&D departments. The series is application-oriented. The level of presentation of each individual volume, however, depends on the subject and can range from practical to scientific.

More information about this series at http://www.springer.com/series/4748

Ashish M. Kothari • Vedvyas Dwivedi
Rohit M. Thanki

Watermarking Techniques for Copyright Protection of Videos

 Springer

Ashish M. Kothari
Atmiya Institute of Technology
and Science
Rajkot, Gujarat, India

Vedvyas Dwivedi
C. U. Shah University
Wadhwan City, Gujarat, India

Rohit M. Thanki
C. U. Shah University
Wadhwan City, Gujarat, India

ISSN 1860-4862   ISSN 1860-4870   (electronic)
Signals and Communication Technology
ISBN 978-3-030-06530-0   ISBN 978-3-319-92837-1   (eBook)
https://doi.org/10.1007/978-3-319-92837-1

Printed on acid-free paper

This Springer imprint is published by the registered company Springer International Publishing AG part of Springer Nature.
The registered company address is: Gewerbestrasse 11, 6330 Cham, Switzerland

Preface

Digital watermarking is a technique to put a secret message, which may be logo of a company, name of the creator, etc., behind the cover medium, which may be image, audio, or video. Digital watermarking algorithms should fulfill three requirements of robustness, perceptibility, and payload capacity. Digital watermarking algorithms may be designed in spatial domain or transform domain. Watermarking in the spatial domain alters the pixel value directly depending on the watermark and that in the transform domain modifies the frequency coefficients of the cover medium depending on the watermark.

This book explains three of the spatial domain video watermarking algorithms, namely visible watermarking, invisible watermarking using Least Significant Bit (LSB) substitution method, and invisible watermarking using correlation-based approach. Basically visible watermarking is used for giving the identity of the producer of the cover medium that can be made visible on the cover medium. The identity includes transparent and nontransparent parts, and the algorithm would superimpose identity on the cover medium in such a manner that the part of the cover medium which is superimposed by the identity is replaced with the identity and the other part remains unaltered. In the invisible watermarking using LSB algorithm, Most Significant Bit (MSB) of the message is replaced with the LSB of the frame of the video. If the message is smaller than the frame, then multiple copies of the message is placed in the frame so as to make the frame robust against cropping attack. In the third spatial domain method of invisible watermarking, namely correlation-based watermarking, one Pseudo-random Noise (PN) sequence is added to the block of the frame of video depending on the message bit. At the receiver side, the same PN sequence is correlated with the block of frame and the message is recovered based on the amount of correlation achieved. Work continues in the transform domain wherein Discrete Cosine Transform (DCT) and Discrete Wavelet Transform (DWT) are used for invisible watermarking. In the DCT-based method, two pixels are used for embedding the message wherein selections of these pixels are made according to the standard JPEG quantization table. In the DWT-based method the frame first undergoes a sub-band coding and then horizontal sub-band is modified according to the message. One linear algebra method, namely Singular Value Decomposition (SVD), is also used for the purpose of embedding both binary and grayscale watermarks. Here the singular values of the frame of video are modified according to the message.

A hybrid algorithm is also explained wherein DCT, DWT, and SVD are combined so as to have advantages of all three. For the evaluation of perceptibility at the transmitter side, two pixel quality matrices are calculated in each of the above-mentioned methods, namely peak signal-to-noise ratio and mean square error. For the sake of evaluation of the robustness, i.e., quality of the message recovered at receiver end correlation between the recovered message and original message is calculated for each method. Based on the experiments the hybrid method is found to be better in both from the point of view of perceptibility and robustness. Finally, compressive sensing-based watermarking algorithms for copyright protection of digital videos are explained. These are new watermarking techniques and known as sparse domain watermarking.

This book is a Ph.D. research work and extension work of Dr. Ashish Kothari submitted to the Department of Electronics and Communication Engineering, Shri Jagdish Prasad Jhabarmal Tibrewal University (JJTU), Jhunjhunu, Rajasthan, in 2013, under the supervision of Dr. Vedvyas Dwivedi. The authors are indebted to numerous colleagues for valuable suggestions during the entire period of the manuscript preparation. We would also like to thank the publishers at Springer, in particular Mary E. James, senior publishing editor/CS Springer, for their helpful guidance and encouragement during the creation of this book.

Rajkot, Gujarat, India   Ashish M. Kothari
Wadhwan City, Gujarat, India   Vedvyas Dwivedi
                                                                         Rohit M. Thanki

Contents

List of Figures

List of Tables

Chapter 1
Introduction

1.1 Prologue

Today's era is the era of heralded connectivity which means that communication through the Internet and network may be wired or wireless. We do have some extraordinary inventions like digital camera, camcorders, MP3 players, PDAs, etc. for creating, manipulating, and enjoying the multimedia data. Nowadays the development of the Internet has given us some valuable gifts like electronic publishing of various files, e-advertising, e-newspaper, e-magazine, e-library, online video, online audio, online product ordering, online transactions, real-time information delivery, and many more. Because of all these, storing, transmitting, and distributing digital videos over the Internet has been a very easy task. However creators of the videos are afraid of transmitting and distributing these valuable videos because of the problem of copyright protection. It is a very easy task to copy digital data, and when it is paste somewhere, it looks like the original one, and therefore it leads toward the spiteful intent of what is called as piracy.

The best possible way, in which the multimedia data are protected against illegal retransmission and recording, is to put a signal behind the cover medium for the authentication of the owner of the data. The signal is called digital signature or copyright label or watermark. The method is known as digital watermarking [1, 2] which is a state-of-the-art technique to put a secret message behind a cover medium in such a manner that the common man cannot visualize the message with a naked eye and he/she perceives it as a normal cover medium. Message may be the name of the creator, a logo of the company, or any other sign which can be extracted only when some specific algorithm is applied to extract the message, and in this way the proof of ownership can be given. Nowadays the subject of interest is to provide proof of ownership and to prevent unauthorized tempering of the multimedia files. It can easily be done because editing of the files is done digitally. And this is the reason why both industry and academic people are working seriously on digital watermarking. There have been many techniques proposed for multiple areas for the

© Springer International Publishing AG, part of Springer Nature 2019
A. M. Kothari et al., *Watermarking Techniques for Copyright Protection of Videos*, Signals and Communication Technology,
https://doi.org/10.1007/978-3-319-92837-1_1

proof of ownership or copyright protection and also for access control. Robustness, imperceptibility, and payload capacity are important requirements for digital watermarking techniques.

1.2 Basic of Watermarking

Digital watermarking [3, 4] is a state-of-the-art technique to put a secret message behind a cover medium in such a manner that the common man cannot visualize the message with a naked eye and he/she perceives it as a normal cover medium. Message may be the name of the creator, a logo of the company, or any other sign which can be extracted only when some specific algorithm is applied to extract the message, and in this way the proof of ownership can be given. Watermarking also describes the same terminologies as steganography, but there are differences between the steganography and watermarking [3–5].

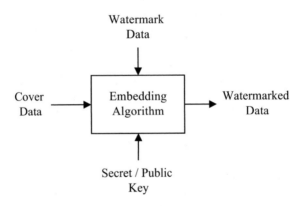

(a) Watermark Embedding Process

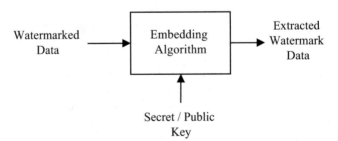

(b) Watermark Extraction Process

Fig. 1.1 Basic watermarking process. (**a**) Watermark embedding process, (**b**) Watermark extraction process

- Steganography can only be used for point-to-point communication, whereas watermarking can be used for point-to-multipoint communication also.
- Steganography methods are very sensitive to any change in the cover medium. If the cover medium is changed, then the hidden message completely vanishes. But in the case of watermarking, there are methods with the help of which one can protect the message even after some degradation given to the cover medium. Technical name given to this phenomenon of keeping the message even after modification of the cover medium is robustness.
- Robustness of the watermarking methods gives more weight age to it compare to steganography.

The process of hiding a watermark is called embedding, and the process of taking the watermark back is called extracting. The ideas of embedding and extraction are given in Fig. 1.1.

1.3   Requirements of Watermarking

There are various requirements [1, 2, 4, 9] that the watermarking algorithm should have. They are explained as under:

- *Robustness*: If the watermarking algorithm is able to preserve the message under various modifications like compression, rotation, cropping, scaling, filtering, or noise addition, then that algorithm is said to be robust.
- *Imperceptibility*: After watermark is embedded in the cover medium, if the visual quality of the cover medium is such that the changes can't be noticed with the naked eye, then the algorithm is called imperceptible.
- *Payload Capacity*: It is the size of the message that can be embedded inside a cover medium. It depends on the algorithm used for the watermarking.

Above three requirements carries a trade-off triangle as shown in Fig. 1.2 which states that to achieve two of the three requirements, third one should be traded off.

For example, to achieve higher robustness and good perceptibility, one needs to throw away the requirement of higher payload capacity. In the same way if more amount of message is to be embedded along with good quality of the picture, the requirement of high robustness must be traded off.

Fig. 1.2 Trade-off between robustness, imperceptibility, and payload capacity

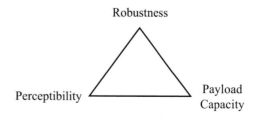

1.3.1 Types of Watermarking

Figure 1.3 shows various ways in which one can use the concept of watermarking [2–4].

- *Spatial Domain Watermarking*: Pixel value of the cover medium is directly changed according to the value of the message to be embedded in this method. The different techniques are defined in the spatial domain watermarking. The examples of spatial domain watermarking algorithms are least significant bit (LSB) and correlation-based algorithm.
- *Transform Domain Watermarking*: Transform domain watermarking implies the use of various transforms to be applied on the cover medium so as to find out the frequency coefficients and then changing these coefficients according to the watermark information. Most powerful transforms used for the purpose of copyright protection are discrete cosine transform (DCT), discrete wavelet transform (DWT), and singular value decomposition (SVD).
- *Invisible Watermarking*: This is the technique of hiding the message in such a way that it can't visibly be detected by a common man.
- *Visible Watermarking*: In this technique the watermark is superimposed on the cover object in such a way that it can be perceptually visible.
- *Source-Based Watermarking*: This technique is used when the owner of a document wants to distribute the document to multiple destinations with the same authentication information. The method is used for authentication purpose. It is possible to detect whether any part of the document is tempered with this method.
- *Destination-Based Watermarking*: The purpose of this kind of technique is the same as source-based scheme, but here each receiver gets unique watermark information that is embedded behind the document. Only that the receiver can open that document. This method can prevent illegal reselling of the document.

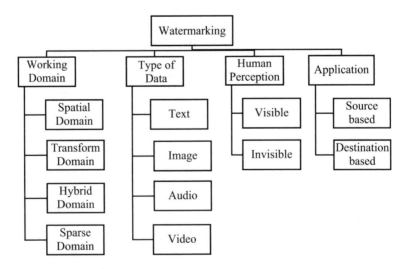

Fig. 1.3 Types of watermarking

1.3.2 Applications of Watermarking

All watermarking schemes are evaluated or compared with respect to either or all of the three requirements, i.e., imperceptibility, robustness, and payload capacity. However without considering the applications [2, 4, 5] of the watermarking, examination of all the requirements would somehow be misleading. The major areas where watermarking is applicable are:

1. Robust identification of digital content
2. Broadcast monitoring of video sequences (digital TV)
3. Advertisement verification
4. Monitoring applications including forensic tracking of distributed video content
5. DVD protection and access control (mastering, adding copyright notices as proof of original ownership)
6. In distribution also known as fingerprinting wherein the operations are addition of copyright notices, to identify recipients and to trace the cause of unlawful copies
7. In remote control and triggering
8. In broadcast chain wherein information are added and hence controlling and triggering of the devices may become possible

1.4 Watermarking Attacks

When a teacher goes to class to deliver a lecture, he/she has to look for the questions that the students may ask. Similarly watermarking algorithm needs to take care of any unwanted weakening to the watermarked medium given by attacks. The main purpose of the watermarking algorithm is to provide some sort of security and that of the attack is to work against this purpose.

When any document is watermarked with a watermark, it may be transmitted in a channel where it may be disturbed intentionally or unintentionally. The owner of the data tries to send the message information as much as possible, and at the same time, he/she takes care of the quality of the cover data. On the opposite end, the work of the attacker is to degrade the quality of the cover data and hence degrade the quality of the watermark. Some of the attacks [8–10] are listed here and are used in this book.

- *Gaussian Low-Pass Filtering*: This is a kind of linear filter. Here in this book, the mask size of the Gaussian filter is made fix at 3×3, and various values of sigma are taken, and performance evolution parameters are measured.
- *Low-Pass Average Filtering*: In this type of filtering, the value of the center pixel value is replaced with the average intensity value of the neighbors. The result of the process is reduced by "sharp" transitions in the gray levels in the image. The

major application of this operation is noise reduction because averaging results in reduction of the sharp transitions. The problem of this kind of filtering is that the process results in the image with blurred edges. Here in this book, various mask sizes for the averaging are taken, and performance evolution parameters are measured.

- *High-Pass Filtering*: The principle objective of linear sharpening is to highlight fine details or enhance details that have been blurred. The sum of the elements of high-pass mask must be equal to zero. This is because when we place the mask on the low-frequency region (i.e., over an area of constant gray level), the result must be zero. The filter reduces the average gray level of the image to zero. It means there are some negative gray levels in the output of the filter. So, some form of scaling is required. The problem with this filter is that only sharp transitions such as edges and boundaries are dominant in the output image and all constant regions are destroyed. Here in this book, all videos undergo this attack and performance evolution parameters are measured.
- *Low-Pass Median Filtering*: This nonlinear filter is also called order statistic filter because the resultant image in the output depends on the ranking or ordering of the values of the pixels within the mask. Here in this book, various mask sizes are taken, and performance evolution parameters are measured.
- *Gaussian Noise Attack*: Gaussian (normal) noise is very attractive from a mathematical point of view since its DFT is another Gaussian process. Gaussian noise PDF is given as

$$p(z) = \frac{1}{\sqrt{2\pi}\sigma} e^{-(z-\mu)^2/2\sigma^2} \tag{1.1}$$

where, z is the intensity, μ is the mean (average) value of z, σ is standard deviation, and σ^2 is the variance of z.

Examples of this noise are electronic circuit noise and sensor noise due to low illumination or high temperature. Here in this book, various mean values are taken, and performance evolution parameters are measured.

- *Impulse/Salt and Pepper Noise Attack*: Noise PDF (bipolar) is given as

$$p(z) = Pa; z = a$$
$$p(z) = Pb; z = b \tag{1.2}$$
$$p(z) = 0; otherwise$$

if $b > a$, intensity b will appear as a light dot on the image and appear as a dark dot. If either Pa and Pb is zero, the noise is called dot. If Pa and Pb is zero, it is called unipolar. Frequently, a and b are saturated values, resulting in positive impulses being white and negative impulses being black. This noise shows up when quick transitions – such as faulty switching – take place. Here in this book, various mean values are taken, and performance evolution parameters are measured.

- *Speckle Noise Attack*: Here in this book, various mean values are taken, and performance evolution parameters are measured.
- *Lossy Compression Attack*: Video is compressed by the method of JPEG compression. Here various quality factors are taken, and performance evolution parameters are measured.
- *Color Reduction Attack*: Color reduction deals with reducing the number of colors from the image. Here in this book 4, 8,16, and 32 color images are taken, and performance evolution parameters are measured.
- *Histogram Equalization Attack*: Histogram of the image is modified so that an equal distribution of the gray level can be done. Performance evaluation parameters are measured after the application of this attack.
- *Horizontal Motion of the Camera Attack*: This is a filter to approximate, and once convolved with a frame it produces the linear motion of a camera by some number of pixels, with an angle of theta degrees in a counterclockwise direction. This generates a vector having both vertical and horizontal motions. In this book various linear motions have been taken keeping the theta constant at 0 degree, and performance evolution parameters are measured.
- *Rotation Attack*: In this book the frame is rotated by various angles in the counterclockwise direction, and performance evolution parameters are measured.
- *Cropping Attack*: In this book various numbers of pixels are cropped, and parameters are measured.

1.5   Quality Measures for Watermarking

The performance of various schemes of watermarking can be evaluated on the bases of some of the visual quality matrices [6, 7, 11, 14] given in eqs. 1.3 and 1.4. These matrices give the idea of visual degradations of the cover medium because of the embedding of the message. These values also give the idea of strength of both watermark and the algorithm. All these matrices depend on the difference created after adding the watermark information and modifying the watermarked documents.

$$\text{MSE} = \frac{1}{MN} \sum_{x=1}^{M} \sum_{y=1}^{N} \left\{ \left(f(x,y) - f'(x,y) \right)^2 \right\} \tag{1.3}$$

$$\text{PSNR} = 10 \times \log \frac{255^2}{\text{MSE}} \tag{1.4}$$

where MSE is a mean square error, PSNR is a peak signal-to-noise ratio, $f(x, y)$ is an original frame of the video, and $f'(x, y)$ is a watermarked frame of the video.

The abbreviation PSNR [11], called peak signal-to-noise ratio, is used to measure the difference and also similarity between a signal in its original form and in its modified form. As per the equations, it is observed that the measurement of PSNR includes the measurement of MSE, mean square error, which actually finds the difference between the signal and modified version of the same. PSNR is measured in the logarithmic scale, and MSE is measured in the general scale. At the extraction side, robustness is measured by finding the correlation between the original watermark image and the recovered watermark image. In this book, PSNR value is used for calculation of imperceptibility quality, and correlation is used for calculation of robustness quality of watermarking approach.

1.6   Colorspace Conversion

Rather than developing the codes in the RGB plain, here the concept of colorspace conversion [25, 34, 35, 37, 39] is used wherein the frame of the video is converted into YCbCr plane. Here Y means the luminance component of the frame, and Cb and Cr are blue-difference and red-difference chrominance components. Here are some of the reasons for choosing YCbCr colorspace:

1. When JPEG compression is performed on the frame RGB, colorspace is affected more as compared to the YCbCr colorspace.
2. It requires less disk space and less bandwidth.
3. This is the only colorspace used in the SD media, which has lower bandwidth and needs to have backward compatibility.

Equations 1.5, 1.6, and 1.7 show the conversion formulas from RGB to YCbCr colorspace, while eqs. 1.8, 1.9, and 1.10 show the reverse conversion.

$$Y = 16 + 128.553 \times G + 65.481 \times R + 24.966 \times B \tag{1.5}$$

$$Cb = 128 - 37.797 \times R + 74.203 \times G + 112 \times B \tag{1.6}$$

$$Cr = 128 - 112 \times R + 93.786 \times G - 18.214 \times B \tag{1.7}$$

$$R = Y + 0 \times Cb + 1.402 \times Cr \tag{1.8}$$

$$G = Y - 0.344136 \times Cb - 0.714136 \times Cr \tag{1.9}$$

$$B = Y + 1.772 \times Cb + 0 \times Cr \tag{1.10}$$

Specifically Y component is selected for the purpose of watermarking because human visual system is more sensitive toward the change in the brightness compare to the color. Therefore JPEG compression does not compress any sample of the Y component. So if message content is embedded in the Y component, it may be better preserved as compared to Cb and Cr. Also even if the frame is compressed, a normal person perceives same information as the original case.

1.7   Related Works to Video Watermarking

Digital video is a sequence of frames which include some important and some redundant signals that are to be stored in hard disc and distributed over some networks. Out of many possible multimedia formats, video is selected over here for the purpose of watermarking. Today's era is the era of storing, transmitting, and distributing digital videos over the Internet. However creators of the video are afraid of transmitting and distributing these valuable videos because of the problem of copyright protection.

Digital watermarking is a state-of-the-art technique to put a secret message behind a cover medium in such a manner that the common man cannot visualize the message with a naked eye and he/she perceives it as a normal cover medium. Message may be the name of the creator, a logo of the company, or any other sign which can be extracted only when some specific algorithm is applied to extract the message, and in this way the proof of ownership can be given.

There are two major domains where the videos are watermarked, namely, spatial domain and transform domain. Spatial domain watermarking provides very high perceptibility such that the quality of the original and watermarked frame is almost the same. Also watermark equal to the size of the frame can be embedded in the frame in the spatial domain. However the problem of spatial domain watermarking is that the robustness achieved is far less comparatively. Transform domain watermarking is far better than spatial domain so far as robustness is concerned and that is the reason why transform domain techniques are preferred.

Inventions in the video watermarking started with embedding messages in the spatial domain. Inventions further carried out in the transform domain with a single transform used for embedding purpose. Further some of the inventors embedded messages with the help of combinations of two transforms and achieved more robustness. The following is the brief summary of the works carried out in this domain of watermarking.

Podilchuk and Wolfgang [1] described the invisible but transparent watermarking scheme for image as well as video. They explained the exploitation of properties of human visual system. They explained the concepts of image and video watermarking using DCT and modification of the frequency coefficients. They also described that more work is required to be performed to make this scheme more robust. They specifically pointed this sentence toward the video watermarking.

Langelaar, Setyawan, and Lagendijk [2] beautifully reviewed the current watermarking techniques, need of watermarking, application of watermarking, and requirement of watermarking algorithms. The authors lightened a lamp on algorithms that has already been implied in spatial and frequency domain. The authors lighted a lamp on what methods and what mathematical expressions are used for changing the pixel value according to the watermark in the spatial as well as in the transform domain. The authors also stated the methods used for the image and video. They also gave the idea of properties of the human visual system.

Podilchuk and Delp [3] have given the idea of what has already been done in the field of watermarking. They explained concepts of digital watermarking. They lighten the lamp on how various multimedia can be watermarked. They covered the concepts of image, audio, video, and graphics watermarking. They also highlighted the work that should be done in the field of watermarking by stating the limitations of the current methods.

Chandramouli, Memon, and Rabbani [4] explained various concepts of digital watermarking. They beautifully highlighted the types of watermarking, applications where digital watermarking may be used, the requirement of watermarking, and difference between the steganography and watermarking.

Paul [5] reviewed the work in the field of watermarking. In her paper, she explained the important aspects of video watermarking, difference between steganography and watermarking, common attacks that can degrade the quality of both video and the message, applications of video watermarking, and techniques in video watermarking wherein she explained the frequency domain video watermarking based on DCT and PCA and spatial domain watermarking based on least significant bit substitution and correlation. She highlighted that the future work can be carried out by cascading the transforms.

Kutter and Petitcolas [6] gave mathematical expressions of visual quality matrices. They gave the expressions for the purpose of comparison and evaluation of robust and invisible watermarking techniques. The authors provided common platform where the inventor of the watermarking algorithm can check the previous method and try to implement a new method which can give better values compared to the previous work. The authors explained various attacks that degrade the multimedia data.

Petitcolas [7] explained the importance to the common platform for the evaluation of the watermarking scheme for the comparison purpose. The author divided the evaluation criteria into two groups. The first group is functionality where evaluation is performed using agreed series of tests. The second group is assurance which includes set of levels to check the first group. The author had given a very good insight to robustness, perceptibility, and capacity.

Voloshynovskiy, Pereira, and Pun [8] said that after watermarking is completed and is transmitted in the channel, there are various operations that may degrade the quality of the cover medium. Possible operations are lossy compression, signal processing algorithms, resampling, etc. They explained properties of watermarking attacks like removal, geometric and protocol attack, estimation-based attacks, and optimized attacks. Authors also highlighted the importance of various benchmarking techniques in evaluating the watermarking techniques.

Huang and Wu [9] highlighted the current visible watermarking schemes specially used for intellectual property rights (IPR) protection. They gave an introduction to some basic requirements of current visible watermarking schemes. Possible security problems and attacking schemes were also included. They mentioned that visible watermarking schemes can be very useful in distributing images and videos with copyrights over Internet.

Khalifa et al. [10] are given an in-depth analysis of types of attacks that can degrade the quality of the watermark. They explained JPEG compression attack, noise attack, rotation and rescaling attacks, blurring attack, contrast enhancement attack in detail, and the effect of the same on the cover medium.

Haroswati et al. [11] used least significant bit (LSB) digital watermarking, and specifically they made use of Halal logo for the authentication purpose. They modified the pixel values by dividing all the bits into most and least significant bits wherein they explained that MSB contains most important information and LSB contains less important information. They explained that these small modifications result in high perceptibility. They generated a QR, Quick Response, code with message for the purpose of embedding Halal logo. They explained concepts of watermarking evaluation parameters, namely, peak signal-to-noise (PSNR), mean square error (MSE), and normalized cross-correlation (NC).

Lee and Chen [12] introduced an image watermarking system where the key part is the variable size LSB insertion. Authors described two different methods where they focused putting as much of bits of watermark in each pixel of the image and achieved a good around 4.02 bits per pixel embedding. In the proposed method, authors made use of two security keys, i.e., stego key and DES key so as to make the watermark content more secured.

Chan and Cheng [13] proposed a spatial domain image watermarking method wherein they used equal-sized grayscale images as the cover image and the secret message. They embedded most significant one to five bits of the secret message in the least significant one to five bits of the cover medium. They explained that if the number of bits to be embedded is increased, perceptibility of the original image reduces and vice versa.

Ramalingam [14] has given the idea of spatial domain video watermarking with a beautiful GUI where one can select the video file to be watermarked along with the key for the embedding purpose. The key is a big text file, and hence it is advantageous so far as security of the key is concerned. The author modified the least significant bit of the video to put the message behind.

Hsu and Wu [15] embedded digital watermarks in the images for the sake of image authentication. They selectively modified the middle frequency parts of the image so as to embed perceptible watermark patterns into the images. They proved the technique is robust against the cropping attack, image enhancement attack, and the compression attack. They also explained that if user key is carefully defined, there may be the possibility of achieving multiple watermarks embedding as well as achieving robustness. They also demonstrated that their technique can equally apply for the images with multiple resolutions by modifying the choice of middle frequencies. The authors failed in achieving the robustness in image resampling and image rotation attacks.

Arena, Caramma, and Lancini [16] explained the technique for MPEG video watermarking where they used the bit stream domain so that the process of data embedding became easy. They embedded a bit of the watermark into a large number of pixels by performing modulation between the watermark and the pseudorandom sequence. They modified I Frame so as to reduce the complexity, and they partitioned the DCT coefficients so as to embed more in higher frequency and less in the lower frequencies. They also suggested the future research points.

Bangaleea and Rughooputh [17] presented a scheme for copyright protection which made use of the spatial domain and attack characterization approach. A small number of bits are embedded onto an image in the spatial domain using a method similar to the direct sequence spread spectrum. Authors described that the imperceptibility of the cover medium and robustness of the watermark data are analyzed using the perceptual analysis wherein the robustness should be increased with perceptibly a very small loss in the cover medium. They added reference and robust watermarks in the algorithm. They checked the robustness against JPEG compression and Gaussian noise attacks.

Lancini, Mapelli, and Tubaro [18] proposed a spatial domain video watermarking scheme on the uncompressed domain and checked the robustness against compression, cropping, and resizing attacks. They used convolution and turbo codes for the improvement in the robustness of the algorithm. They used spread spectrum approach to generate a mask, and mask is added with the original video to get the watermarked video.

He, Sun, and Tian [19] proposed a semi-fragile scheme of video watermarking wherein they made use of discrete Fourier transform to convert the video from the spatial domain to the transform domain. They embedded the watermark bits in the selected angular radical transformation coefficients. They also made use of error-correcting codes for the purpose of watermark generation and embedding. They also suggested some changes required for the better performance.

Hernández at el [20]. suggested a transform domain watermarking technique for the purpose of copyright protection of digital images wherein they made use of a popular transform, namely, discrete cosine transform (DCT) domain. Authors divided the image into blocks of 8x8 and then applied DCT to individual blocks and then changed the frequency coefficients according to the message using spread spectrum sequence. They made use of likelihood estimation function in the proposed method. They experimented the work with JPEG compression and Gaussian noise attacks. They found difficulties in achieving good results against other attacks.

Lu and Liao [21] claimed a proposed scheme to be the first one that is robust toward the geometric attacks like blurring, filtering, scaling, MPEG compression, etc. They explained that in video, the problem of synchronization caused by rotation and flipping is solved by making use of eigenvectors. They also made use of the discrete cosine transform for converting spatial video into the transform domain and then arranged them in a zigzag fashion so as to differentiate the frequency component in low, medium, and high band.

Preda and Vizireanu [22] explained a blind video watermarking scheme making use of the discrete cosine transform. They tested watermarked video with various bit rates and hamming and Reed-Solomon codes. They tested the algorithm with different scenes where different redundancy elements were present. They proved from the results of bit error rate that block quantization method gives promising results both in the spatial and DCT domain. They tested the scheme under compression attack.

Koz and Alatan [23] suggested an oblivious video watermarking technique where they embedded the maximum energy of the watermark and also took care of the perceptibility of the video. They explained the concepts of temporal redundancy and gave its importance of the same in context to the human visual system. They first determine the temporal contrast threshold so as to determine maximum watermark energy that can be embedded, and then they embedded the watermark accordingly.

Sridevi et al. [24] described a discrete cosine transform-based digital video watermarking scheme for MPEG video where in the message to be embedded is a grayscale image. Watermark is first preprocessed using DCT, normalization, and frequency masking. They claimed that because of this preprocessing, the capacity of the video watermarking is greatly increased.

Ding et al. [25] described one of the unique methods to embed a watermark behind the video. They made use of the video features to embed the watermark information where they found the distance between adjacent key frames and made them as watermark information. Then they modified the DC coefficient of the DCT coefficients according to this information. They proved that this technique is robust especially for the copy attack.

Ejima and Miyazaki [26] proposed image and video watermarking scheme whose heart is sub-band coding and specifically discrete wavelet transform. Authors used the colorspace conversion method to convert the RGB video into YCbCr video where Y is the luminance part, and it is most important so far as the human visual system is concerned. They tested the scheme under watermark against image compression, adding noise and smoothing attacks. They embedded the watermark in the Y part taking care that the other two components would be lost when compression is carried out. For more robustness the author suggested to have as much sub-bands as possible.

Serdean et al. [27] described the use of one of the most powerful transform called discrete wavelet transform in the video watermarking wherein they also combined log-log and log-polar transform. The authors made use of speeded PN sequence for the purpose of watermark embedding. In this method authors used two keys to achieve better security. Authors claimed that the method is robust against rotation of 70 degrees, scaling up to 180 degrees, and aspect ratio changes up to 200.

Liang Fan, Fang Yanmei [28] suggested the method based on discrete wavelet transform and gold sequences for the embedding purpose. In the embedding side, they encoded the watermark using CDMA and then embedded those into the lowest frequency band of the DWT transformed frame of video sequence. In the decoding side, the same gold sequence is generated, and the lowest frequency of the watermarked frame is auto-correlated with the same, and hence the watermark is detected. They achieved good visual quality and robustness against compression, noise, cropping, and frame dropping.

Elbasi [29] proposed a semi-blind video watermarking scheme where in a pseudorandom sequence based on the watermark is embedded in all frequency components higher than a threshold. In the extraction side, another threshold is selected, and the frequency coefficients which are higher than the threshold are selected for the detection process which is then correlated with the pseudorandom sequence, and hence the watermark is detected. The author also extended this scheme to embed the watermark in more than one band.

Essaouabi and Ibnelhaj [30] proposed a blind video watermarking scheme which is based on video scene segmentation and 3-D wavelet transform. In this authors first applied the discrete wavelet transform to the grayscale image and decomposed it into four bands. Similarly they decomposed the video frames into four bands and modified the bands according to the respective four bands of the message. They proved the method is robust against frame dropping, frame averaging, and lossy compression.

Raghavendra and Chetan [31] proposed a very novel method for blind video watermarking based on discrete wavelet transform. In this authors divided the binary watermark into four different parts and then embedded each part into the sub-bands of the video frames. Depending on the message bit, they exchanged the current pixel value with maximum or minimum of five different members. Authors compared the scheme with the existing DWT-based scheme and found better results in terms of robustness against frame-dropping problem.

Hussein and Mohammed [32] proposed a robust method using discrete wavelet transform and motion estimation algorithm. Authors choose horizontal detail component and vertical detail component of the video frame so as to embed the watermark because the motion in these bands does not affect the quality of the extracted watermark. Authors embedded the watermark using an addition of the pseudorandom sequence with the original band.

Mostafa et al. [33] combined the discrete wavelet transform and principal component analysis for the purpose of video watermarking. In the embedding side, first the original video frame is decomposed into four sub-bands, and then the block-based PCA is applied to one of the bands. The result is added with the watermark frame, and inverse DWT would give the watermarked frame. In the extraction side, correlation property is used so as to detect the watermark.

Khatib, Haj, and Rajab [34] beautifully made use of one of the most important linear algebra tool, i.e., singular value decomposition. They explained the embedding of a watermark in all three matrices, namely, two unitary, i.e., U and V, and one singular matrix. They first performed the colorspace conversion, and then they applied SVD on the Y matrix and changed the U, V, or S values according to the watermark. They proved this method is more robust against attacks.

Kamlakar, Gosavi, and Patankar [35] explained a video watermarking method based on block-based SVD. They first explained the advantages of the singular value decomposition and then embedded the watermark by modifying the singular values of the original frame according to that of the watermark. They first divided the frames in the blocks and selected one of the blocks for the embedding purpose. They explained this method to be robust against the compression, rotation, noise, translation, and scaling attack.

Huang and Guan [36] combined DCT and SVD so as to embed the watermark message into the cover medium. Moreover they used a method called LPSNR wherein they set the value of the PSNR, and accordingly they embedded the grayscale watermark into the cover medium by modifying the singular values of the DCT coefficient of the cover medium. They checked and verified the robustness of the method against many attacks including rotation, scaling, cropping, and noise attacks.

Rajab, Khatib, and Haj [37] combined DWT and SVD for the purpose of putting the watermark behind a digital video. Authors first used the colorspace conversion and used Y frame of the video for the watermarking purpose. They decomposed the Y frame into seven sub-bands and used the horizontal detail sub-band for the watermarking purpose. SVD is applied on the sub-band, and singular values are modified according to that of the watermark.

Mansouri et al. [38] described the method of watermarking where they first applied complex wavelet transforms to both the frame and the watermark. Then they selected one of the sub-bands of the frame and the watermark and applied SVD to them. They changed the singular values of the sub-band of the original frame according to the singular values of the sub-band of the watermark. Santhi and Thangavelu [39] explained that embedding the watermark into the RGB colorspace is not suggestible because information in R, G, and B color space are highly correlated, and therefore they suggested YUV colorspace for putting the message behind the cover medium. They explained the concept of combining the SVD and DWT for the purpose of watermarking. They found the method is robust against noise attacks and histogram equalization attack.

1.8 Book Organization

This chapter briefly discussed general characteristics of biometric system and multibiometric system. In addition, the motivation behind writing this book is presented. The rest of this book is organized as follows. Chapter 2 explains the concepts of spatial domain watermarking wherein visible and invisible watermarking is explained. Invisible watermarking is implemented in two different ways, one using the least significant bit substitution method and the other using the correlation-based method. Chapter 3 explains another method to embed binary message behind video, namely, transform domain watermarking wherein rather than modifying the pixel values of the cover video, frequency coefficients are modified according to the message. In the chapter two of the most powerful transform are discrete cosine transform (DCT) and discrete wavelet transform (DWT). Chapter 4 explains singular value decomposition (SVD)-based approaches for embedding of different types of messages into video frame. In Chap. 5, a novel method of watermarking is explained wherein all three powerful tools, i.e., DCT, DWT, and SVD, used in previous chapters, are combined for the purpose of watermarking. It is proved in this chapter that this method is the best so far as both perceptibility and robustness are concerned. Chapter 6 demonstrates video watermarking algorithm using compressive sensing (CS) theory procedure. Chapter 7 gives discussion on results obtained from all presented watermarking techniques and some future direction.

Bibliography

1. R.B. Wolfgang, C.I. Podilchuk, Perceptual watermarks for digital images and video. Proc. IEEE **87**(7), 1108–1126 (1999)
2. G.C. Langelaar, I. Setyawan, R.L. Lagendijk, Watermarking of digital image and video data – A state of art review. IEEE. Signal. Process. Mag. **17**(5), 20–46 (2000)
3. C.I. Podilchuk, E.J. Delp, Digital watermarking: Algorithms and applications. IEEE Signal Process. Mag. **18**(4), 33–46 (2001)
4. R. Chandramouli, N.D. Memon, M. Rabbani, Digital watermarking, in *Encyclopedia of Imaging Science and Technology*, (Wiley, New York, NJ, 2002)
5. R.T. Paul, in *Review of Robust Video Watermarking Techniques*, IJCA Special Issue on Computational Science – New Dimensions & Perspectives. (NCCSE, 2011), Cochin, Kerala, India
6. M. Kutter, F.A.P. Petitcolas, A fair benchmark for image watermarking systems. Electron. Imaging Secur. Watermarking Multimed. Contents **3657**, 25–27 (1999)
7. F.A.P. Petitcolas, Watermarking schemes evaluation. IEEE Signal Process. Mag. **17**, 58–64 (2000)
8. S. Voloshynovskiy, S. Pereira, T. Pun, Attacks on digital watermarks: Classification, estimation-based attacks, and benchmarks. IEEE Commun. Mag. **39**, 118–126 (2001)
9. C.-H. Huang., J.-L.Wu Attacking visible watermarking schemes. IEEE Trans. Multimed. **6**(1), 16–30 (2004)
10. O.O. Khalifa, Y. binti Yusof, A.-H. Abdalla and R.F. Olanrewaju, State-of-the-art digital watermarking attacks. Int. Conf. Comput. Commun. Eng. (ICCCE). 744–750 (2012)
11. CKHCK. Yahaya, H. Hassan and MIBM. Kahmi, Investigation on perceptual and robustness of LSB digital watermarking scheme on Halal Logo authentication. Proc. Int. Conf. Syst. Eng. Technol. 11–12 Sept (2012)
12. Y.K. Lee, L.H. Chen, High capacity image steganographic model. IEEE Proc. Vis. Image Signal Proces. **147**, 288–294 (2000)
13. C.-K. Chan, L.M. Cheng, Hiding data in images by simple LSB substitution. Pattern Recogn. **37**, 469–474 (2004)
14. M. Ramalingam, Stego machine – Video steganography using modified LSB algorithm. World Acad. Sci. Eng. Technol. **74**, 502–505 (2011)
15. C.-T. Hsu, J.-L. Wu, Hidden digital watermarks in images. IEEE Trans. Image Proces. **8**, 58–68 (1999)
16. S. Arena, M. Caramma, R. Lancini, Digital watermarking applied to MPEG-2 coded video sequences exploiting space and frequency masking. Proc. Int. Conf. Image Proces. **2**, 796–799 (2000)
17. R. Bangaleea and H.C.S. Rughooth, Performance improvement of spread Spectrum spatial domain watermarking scheme through diversity and attack characterization. in IEEE conf. Africon. 293–298 (2002)
18. R. Lancini, F. Mapelli, S. Tubaro, A robust video watermarking technique in the spatial domain. IEEE Region 8 Int. Symp. Video/Image Proc. Multimed. Commun. Zadar, Croatia. 251–256, 16–19 June 2002
19. D. He, Qibin Sun and Qi Tian, A semi-fragile object based video authentication system. Proc. Int. Symp. Circuits Syst. 814–817 (2003)
20. J.R. Hernandez, M. Amado, F. Perez-Gonzalez, DCT-domain watermarking techniques for still image: Detector performance analysis and a new structure. IEEE Trans. Image Proces. **9**, 55–68 (2000)
21. L. Chun-Shien, H.-Y.M. Liao, Video object-based watermarking: A rotation and flipping resilient scheme. Proc. Int. Conf. Image Proc. (2001)
22. R. O. Preda, D. N. Vizireanu, Blind watermarking capacity analysis of MPEG2 coded video, Proc. Conf. Telecommun. Modern Satell. Cable Broadcast. Serv. Serbia. 465–468 (2007)

23. A. Alper Koz, A. Alatan, Oblivious Spatio-Temporal Watermarking of Digital Video by Exploiting the Human Visual System. IEEE Trans Circuits Syst. Video Technol. **18**(3), 326–337 (2008)
24. T. Sridevi, B. Krishnaveni, V. Vijaya Kumar, Y. Rama Devi, A video watermarking algorithm for MPEG videos, A2CWiC 2010 – Amrita ACM-W celebration of women in computing, 16–17 Sept 2010
25. Y. Ding, X. Zheng, Y. Zhao, G. Liu, A video watermarking algorithm resistant to copy attack, Proc. Third Int. Symp. Electron. Commerce Secur. 29 July 2010
26. Masataka ejima, Akio Miyazaki, A wavelet-based watermarking for digital images and video, IEEE Int. Conf. Image Proc. 678–681 (2000)
27. C.V. Serdean, M.A. Ambroze, M. Tomlinson and G. Wade, Combating geometrical attacks in a dwt based blind video watermarking system. IEEE Region 8 Int. Symp. Video/Image Proces. Multimed. Commun. Zadar, Croatia, 263–266, 16–19 June 2002
28. F. Liang, F. Yanmei, A DWT-based video watermarking algorithm applying DS-CDMA. IEEE Region 10 Conf. TENCON 2006. 14–17 Nov 2006
29. E.L.B.A.S.I. Ersin, Robust mpeg video watermarking in wavelet domain. Trakya Univ. J. Sci. **8**(2), 87–93 (2007)
30. A. Essaouabi, E. Ibnelhaj, A 3D wavelet -based method for digital video watermarking. Proc. 4th IEEE Int. Inf. Hiding Multimed. Signal Proc. 29–31 July 2009
31. Raghavendra K, Chetan K.R, A blind and robust watermarking scheme with scrambled watermark for video authentication. Proc. IEEE Int. Conf. Internet Multimed. Services Arch. Appl. 9–11 Dec 2009
32. J. Hussein, A. Mohammed, Robust video watermarking using multi-band wavelet transform. IJCSI Int. J. Comput. Sci. Issues **6**(1), 44–49 (2009)
33. S.A.K. Mostafa, A.S. Tolba, F.M. Abdelkader, H.M. Elhindy, Video watermarking scheme based on principal component analysis and wavelet transform. IJCSNS Int. J. Comput. Sci. Netw. Secur. **9**(8), 45–52 (2009)
34. T. Al-Khatib, A. Al-Haj, L. Rajab, Video watermarking algorithms using the SVD transform. Eur. J. Sci Res **30**(3), 389–401 (2009)
35. Meenal A. Kamlakar, Chhaya Gosavi, Abhijit J. Patankar, Single channel watermarking for video using block based svd. Int. J. Adv. Comput. Inf. Res. **1**(2), (2012)
36. F. Huang, Z.-H. Guan, A hybrid SVD-DCT watermarking method based on LPSNR. Pattern Recogn. Lett. **25**, 1769–1775 (2004)
37. L. Rajab, T. Al-Khatib, A. Al-Haj, Hybrid DWT-SVD Video Watermarking. Proc. Int. Conf. Innov. Inf. Technol. 16–18 Dec 2008
38. A. Mansouri, A. Mahmoudi, F. Aznaveh, T. Azar, SVD-based digital image watermarking using complex wavelet transform. Sadhana **34**(3), 393–406 (2009)
39. V. Santhi, Arunkumar Thangavelu, DWT-SVD combined full band robust watermarking technique for color images in YUV color space. Int. J. Comput. Theory Eng. **1**(4), 424–429 (2009)

Chapter 2
Video Watermarking in Spatial Domain

Most of the early researches in the video watermarking evolved the use of changing the pixel values of frames of the video directly depending on the characteristics of the digital watermark. These methods are straightforward and less computationally expensive. There are two types of video watermarking methods that can be carried out in spatial domain: (1) visible watermarking and (2) invisible watermarking.

2.1 Visible Watermarking

Here visible watermarking [1, 2] means that the message to be embedded is put in such a manner that it appears perceptibly visible to everyone. This method is used for the authentication and ownership of the cover data. Let us assume that the video frame contains M number of rows and N number of columns. Here the visible message W has the same size as X. Moreover, this visible watermark, as shown in Fig. 2.1, can be divided into two parts:

1. The nontransparent part W_{NT} (the visible information of W)
2. The transparent part WT (the transparent background of W)

The nontransparent part of W is the watermark pattern itself. It can be easily recognized on X. But part of the frame does not get changed where the transparent part is present on W. In the actual algorithm, visible watermark is embedded using the following steps:

Step 1. Original video is converted into frames.
Step 2. Frames are divided into its red, green, and blue planes.
Step 3. Each plane is taken subsequently, and steps 4 and 5 are performed on each plane.

Fig. 2.1 Visible watermark

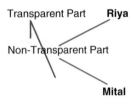

Transparent Part   **Riya**

Non-Transparent Part

Mital

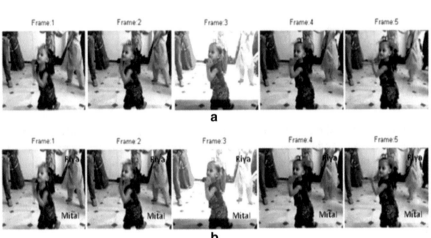

Fig. 2.2 Example of visible video watermarking. (**a**) Original video frames, (**b**) Watermarked video frames

Step 4. When there is a black pixel (00000000), i.e., the nontransparent part of the message, the corresponding pixel of the plane is replaced with the black pixel so as to have a visual perception.

Step 5. When there is a white pixel (11111111), i.e., the transparent part of the message, the corresponding pixel of the plane is kept as it is.

Step 6. Watermarked planes are combined, and the watermarked frame is obtained.

Step 7. Steps 2 to 6 are executed for the next frame, and the process continues until the last frame.

Step 8. Visible watermarked video is obtained by combining all watermarked frames.

Figure 2.2 shows the visible watermarking process on one of the digital videos. In this book, five frames of videos are used for demonstration of experimental results for better visualization of results.

2.2 Invisible Watermarking

Invisible watermarking means putting the digital watermark behind the cover object in such a manner that it cannot be perceptibly visible to all. Only the knowledge of the key and decoding algorithm would help getting the message back from the watermarked video. This method is used for giving the proof of ownership of the cover data. Basically two major ways to perform the task in the spatial domain are:

1. Least significant bit substitution approach
2. Correlation-Based Approach

2.2.1   Least Significant Bit Substitution Approach

In least significant bit substitution method [3–5], most significant bit/bits of the watermark is/are replaced with the least significant bit/bits of the cover data. Here visual degradation is very less because the less important part of the cover medium is affected. Consider an M × N image where M shows the number of rows and N shows the number of columns. In this image a particular pixel is having an intensity value which is actually a decimal value in between 0 and L-1 where $L = 2^k$ and k = number of bits used to represent a pixel. In an 8-bit grayscale image, values are in between 0 and 255, and each positive number β_{10} can be represented by

$$\beta_{10} = b0 + b1 \times G1 + b2 \times G2 + \dots \qquad (2.1)$$

where $G = 2$. Figure 2.3 shows LSB decomposition of a pixel of an image. Here the procedure of decomposition allows the image to be broken into 8-bit planes.

The most significant bit (i.e., Bit 7) of the pixel contains the most important and majority of the visually significant data, while the least significant bit (i.e., Bit 0) contains no visually important data. All other bit planes contribute to the more subtle details in a pixel. The above sentence can be visually elaborated in Fig. 2.4 where bit plane representation of an image is given. It can be observed that the most important information lies in the MSB of the image, and going from MSB to LSB, the amount of information starts degrading. It can be seen that LSB does not contain any important information.

Fig. 2.3 LSB decomposition of one pixel

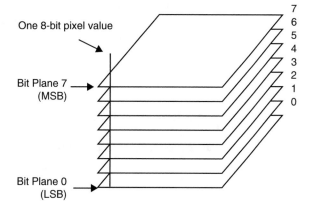

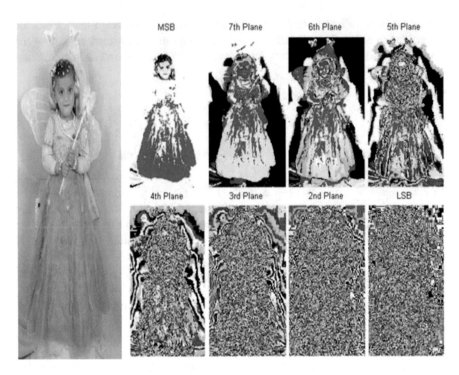

Fig. 2.4 Original image and its bit representation

2.2.1.1   Watermark Embedding Process

The process of embedding a message in the video using this method involves the following steps:

Step 1.   Original video is broken into a number of frames.
Step 2.   Message is repeated and scaled according to the size of the frame.
Step 3.   Frames are divided into its R, G, and B planes.
Step 4.   Any of the three planes is selected for the embedding purpose.
Step 5.   LSB of each pixel of the frame is replaced with MSB of the message, and the result is a frame with watermark embedded in it.
Step 6.   Steps 2 to 5 are executed for the next frame, and the process continues until the last frame.
Step 7.   Watermarked video is obtained by combining all watermarked frames.

2.2.1.2   Watermark Extraction Process

Since the watermark is embedded in the LSB of the frame, the extraction process involves extracting LSB from each pixel of the frame and copying that LSB to all other planes. Stepwise implementation of the extraction process is given as follows:

Step 1.   Watermarked video is broken into a number of frames.
Step 2.   Frames are divided into its R, G, and B planes.
Step 3.   Plane used in the embedding is selected for the extraction purpose.
Step 4.   qLSB of each pixel of the frame is taken and copied into other seven planes.
Step 5.   Steps 2 to 4 are executed for the next frame, and the process continues until the last frame.

2.2.1.3   Experimental Results

Here five different messages have been taken for embedding in the five different video frames as shown in Fig. 2.5.

Bit plane representation of first watermarked frame is shown in Fig. 2.6, and implementation of the above procedure on video is depicted in Fig. 2.7.

Figure 2.8 shows the extracted messages from the first five watermarked video frames.

The quality measures for LSB substitution approach are summarized in the Table 2.1.

Fig. 2.5 Messages to be embedded

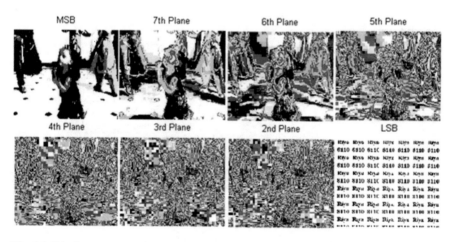

Fig. 2.6 Bit plane representation of first watermarked video frame

Fig. 2.7 Example of invisible video watermarking using LSB substitution approach. (a) Original video frames, (b) Watermarked video frames

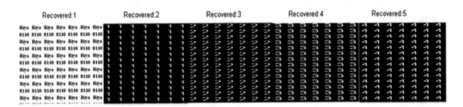

Fig. 2.8 Extracted messages using LSB substitution approach

Table 2.1 Quality measure values for LSB substitution approach

Video frame	PSNR (dB)	Correlation
Frame 1	64.3478	1
Frame 2	56.4755	1
Frame 3	55.2216	1
Frame 4	57.0536	1
Frame 5	56.6812	1

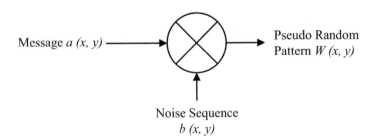

Fig. 2.9 Generation of pseudorandom sequence

2.2.1.4   Observation on Obtained Results

The main advantage of the LSB technique is that after embedding the message into the frame, the maximum variation in the intensity value of the frame is only one. This would not affect the quality of the frame as perceived by human. Hence under the normal circumstances, an average human cannot detect the embedding of a message behind the frame, and hence the transfer of message remains unnoticed or imperceptible to the average human.

From the extracted watermark, one can observe that this method is very powerful when subjected to cropping attack because even if most of the multiple watermarks are lost while the watermarked frame undergoes this attack, the retrieval of a single watermark would be considered as a success.

From the discussions done and results from the prior art, it can be observed that this method is not capable of extracting the message effectively when the watermark frame undergoes certain attacks other than cropping.

2.2.2   Correlation-Based Approach

Another possibility of embedding the digital watermark in the spatial domain is to exploit the correlation properties [6–9] of the noise pseudorandom patterns which are additive in nature. These patterns are utilized for the purpose of watermarking because they carry very good characteristic of low amplitude like noise, do have great correlation property, and are resistant to interference. PN sequences are utilized for the purpose of watermarking due to the reasons as given below:

1. They are random.
2. An initial seed is used to generate them.
3. They are passed through many steps of randomness.
4. It becomes very difficult to predict this sequence until and unless there is knowledge of the seed as well as the algorithm.

In the proposed algorithm, two pseudorandom sequences are generated using the same key. One will be used when the watermark is logic 1 and the other is used when it is logic 0. Figure 2.9 shows the idea of generation of the pseudorandom pattern.

2.2.2.1   Watermark Embedding Process

The following is the sequential procedure used for the purpose of embedding a digital watermark into the video.

Step 1.   Original video is broken into a number of frames.
Step 2.   Block size is decided based on the size of the frame, and gain factor is selected.

Step 3.   Two highly uncorrelated PN sequences are generated using the key.
Step 4.   If message bit contains zero, PN sequence zero is added to that portion of
          watermark mask. Otherwise mask is filled with PN sequence one.
Step 5.   Add watermark mask to frame using gain factor K. Eq. 2.2 shows the
          process.

$$Wf(x,y) = f(x,y) + k \times W(x,y) \tag{2.2}$$

where $Wf(x, y)$ is a pixel of watermarked video frame, $f(x, y)$ is a pixel of original
video frame, $W(x, y)$ is a pseudorandom noise (PN) pattern, and k is a gain factor.
Step 6.   Repeat the process until all the frames are watermarked.
Step 7.   Watermarked video is obtained by combining all watermarked frames.

2.2.2.2   Watermark Extraction Process

The following is the sequential procedure used for the purpose of extracting and
embedding a digital watermark from the video.
Step 1.   Watermarked video is broken into a number of frames.
Step 2.   Block size is selected.
Step 3.   Two highly uncorrelated PN sequences are generated using the key.
Step 4.   Frame is divided into a set of blocks.
Step 5.   Correlation of each block with each PN sequence is calculated.
Step 6.   If correlation with PN sequence one is higher than that with PN sequence
          zero, then make the message bit 1. Otherwise make it 0. Perform this
          operation for all blocks and get the message frame.
Step 7.   Take the next frame, and repeat steps 4 to 6 until all frames are completed.

2.2.2.3   Results of Imperceptibility Test

For imperceptibility test of correlation-based approach, the cover digital video with a
size of 512 × 512 pixels and monochrome messages with a size of 64 × 64 pixels are
taken. Here, video frame is divided into 8 × 8 non-overlapping blocks. The one bit of
message is embedded into each block of video frame to the generated watermarked
frame using gain factor. Figure 2.10 shows the results of correlation-based approach
with gain factor 100 and block size 8. The quality measures for correlation-based
approach for gain factor 100 and block size 8 are summarized in Table 2.1 and 2.2.

In the watermark embedding process, the gain factor is playing an important
role. The visual quality of watermarked video and extracted messages is depending
on the value of the gain factor. Figure 2.11 shows results of the method on Frame 1
considering various values of the gain factor. Table 2.3 shows the results of
watermarking Frame 1 with various gain factors.

Fig. 2.10 Results of imperceptibility test of correlation-based approach. (a) Original video frames, (b) Watermarked video frames, (c) Extracted messages

Table 2.2 Quality measure values for correlation-based approach

Video frame	PSNR (dB)	Correlation
Frame 1	29.1594	0.9157
Frame 2	29.2581	0.9139
Frame 3	27.7918	0.9124
Frame 4	29.7354	0.9263
Frame 5	29.2402	0.9192

2.2.2.4   Results of Robustness Test

For robustness test, various types of watermarking attacks are applied on watermarked video frames, and messages are extracted from these corrupted watermarked video frames. If the message is successfully extracted from the corrupted frames, then the approach is said to be robust. Here various types of watermarking attacks such as filtering attacks, addition of noise, geometric attacks, and other signal processing with different values are applied on watermarked video frame.

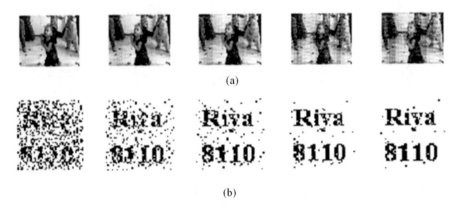

(a)

(b)

Fig. 2.11 Results of correlation-based approach with various gain factor values like 10, 30, 50, 70, and 90. (**a**) Watermark video frame 1, (**b**) Extracted messages

Table 2.3 Results of correlation-based approach on Frame 1 using various gain factors

Gain factor (k)	PSNR (dB)	Correlation
10	39.2247	0.3571
30	32.0259	0.6411
50	30.4713	0.8025
70	29.6605	0.8781
90	29.2698	0.9068
100	29.1594	0.9157

Figures 2.12, 2.13, 2.14, and 2.15 show the results of robustness test for correlation-based approach against different filters such as average filter, Gaussian low-pass filter median filter, and high-pass filter, respectively. Here the numbers on the upper side of the video frames show PSNR values and that in the lower side of the frames show variants of particular attacks. Similarly the numbers on the upper side of the recovered watermark show correlation values.

The compression is very basic attacks which affect any multimedia data when it is transmitted over the communication channel. Figure 2.16 shows the results of robustness test for correlation-based approach against JPEG compression with various quality factors. Figures 2.17, 2.18, and 2.19 show the results of robustness test for correlation-based approach against attacks such as color reduction, histogram equalization, and motion blurring.

Figures 2.20, 2.21, and 2.22 show the results of robustness test for correlation-based approach against various noise addition attacks such as Gaussian noise, salt and pepper noise, and speckle noise. Figures 2.23 and 2.24 show the results of robustness test for correlation-based approach against geometric attacks such as rotation and cropping.

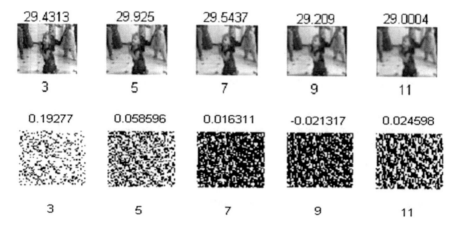

Fig. 2.12 Watermarked video frame and extracted message for correlation-based approach under average filtering attack with various mask sizes

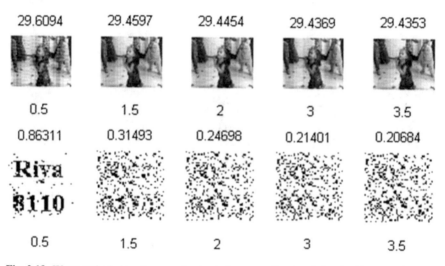

Fig. 2.13 Watermarked video frame and extracted message for correlation-based approach under Gaussian low-pass filtering attack with various standard deviations

2.2.2.5 Observation on Obtained Results

The following are some of the observations made after successfully implementing both embedding and extracting algorithm. Here gain factor of 100 is assumed for the sake of observations in terms of perceptibility and robustness. The higher is the value of PSNR, the higher is the perceptibility, and the higher is the value of correlation, the higher is the robustness.

- Perceptibility in this approach decreases with the increase in gain factor.
- Robustness increases with the increase in gain factor.
- The frames look visibly fine if the resultant PSNR is above 28 dB. The message seems visibly identifiable if the resultant correlation is greater than 0.50.
- This method is not robust against average filtering, median filtering, rotation, and high-pass filtering attacks.
- This method is partially robust against Gaussian low-pass filtering, compression, linear motion of camera, Gaussian noise, salt and pepper noise, and speckle noise attacks.
- This method is fully robust against color reduction and cropping attacks.

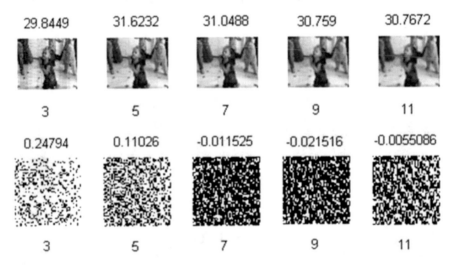

Fig. 2.14 Watermarked video frame and extracted message for correlation-based approach under median filtering attack with various mask sizes

Fig. 2.15 Watermarked video frame and extracted message for correlation-based approach under high-pass filtering attack

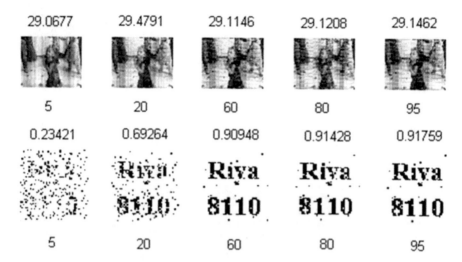

Fig. 2.16 Watermarked video frame and extracted message for correlation-based approach under compression attack with various quality values

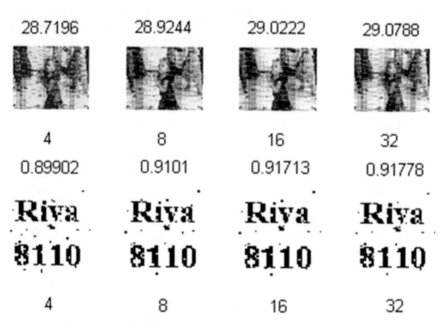

Fig. 2.17 Watermarked video frame and extracted message for correlation-based approach under color reduction attack with various number of colors

Fig. 2.18 Watermarked video frame and extracted message for correlation-based approach under histogram equalization attack

28.1026

0.91759

30.0336 29.1435 28.8412 28.6023 28.3972

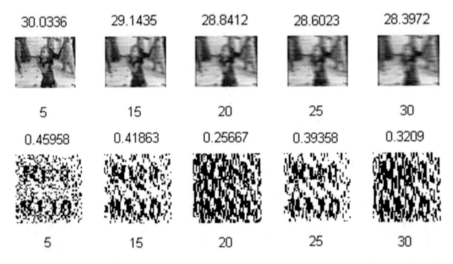

5 15 20 25 30

0.45958 0.41863 0.25667 0.39358 0.3209

5 15 20 25 30

Fig. 2.19 Watermarked video frame and extracted message for correlation-based approach under motion blurring attack

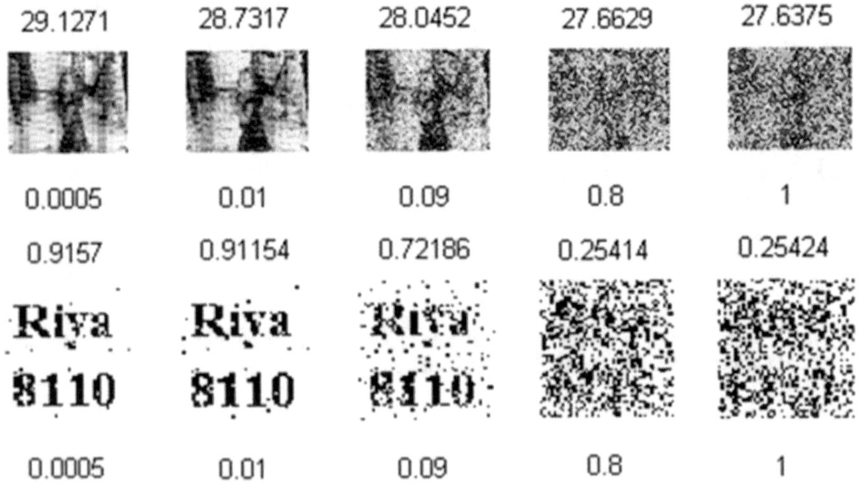

29.1271 28.7317 28.0452 27.6629 27.6375

0.0005 0.01 0.09 0.8 1

0.9157 0.91154 0.72186 0.25414 0.25424

0.0005 0.01 0.09 0.8 1

Fig. 2.20 Watermarked video frame and extracted message for correlation-based approach under Gaussian noise attack with zero mean and various variances

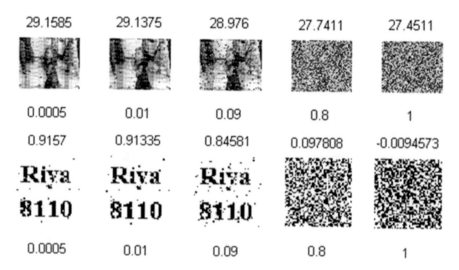

Fig. 2.21 Watermarked video frame and extracted message for correlation-based approach under salt and pepper noise attack with various variances

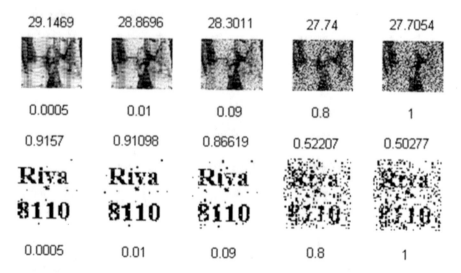

Fig. 2.22 Watermarked video frame and extracted message for correlation-based approach under speckle noise attack with various variances

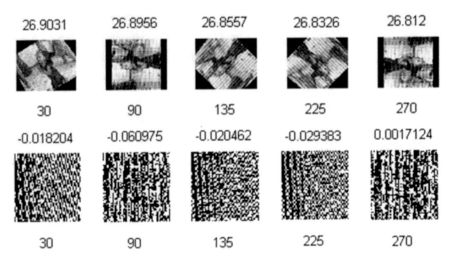

Fig. 2.23 Watermarked video frame and extracted message for correlation-based approach under rotation attack with various angles

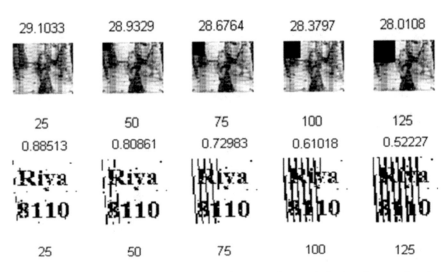

Fig. 2.24 Watermarked video frame and extracted message for correlation-based approach under cropping attack with various crop regions

Bibliography

1. C.-H. Huang, J.-L. Wu, Attacking visible watermarking schemes. IEEE Trans. Multimed. **6**(1), 16–30 (2004)
2. CKHCK. Yahaya, H. Hassan and MIBM. Kahmi, Investigation on perceptual and robustness of LSB digital watermarking scheme on Halal Logo authentication. Proceedings of International Conference on System Engineering and Technology 11-12 September 2012
3. Y.K. Lee, L.H. Chen, High capacity image steganographic model. IEEE proceedings of Vision Image and signal processing, 288–294 (2000)
4. C.-K. Chan, L.M. Cheng, Hiding data in images by simple LSB substitution. Pattern Recogn. **37**, 469–474 (2004)
5. M. Ramalingam, Stego machine – Video steganography using modified LSB algorithm. World Acad. Sci. Eng. Technol. **74**, 502–505 (2011)
6. C.-T. Hsu, J.-L. Wu, Hidden digital watermarks in images. IEEE Trans. Image Proces **8**, 58–68 (1999)
7. S. Arena, M. Caramma, R. Lancini, Digital watermarking applied to MPEG-2 coded video sequences exploiting space and frequency masking. Proc. Int. Conf. Image Proces. **2**, 796–799 (2000)
8. R. Bangaleea and H.C.S. Rughoopth, Performance improvement of spread Spectrum spatial domain watermarking scheme through diversity and attack characterization. in IEEE conf. Africon, 293–298, (2002)
9. R. Lancini, F. Mapelli, S. Tubaro, A robust video watermarking technique in the spatial domain. IEEE Region 8 Int. Symp. Video/Image Proces.Multimed. Commun. Zadar, Croatia, 251–256, 16–19 June 2002

Chapter 3
Video Watermarking in Transform Domain

Transform domain video watermarking algorithm [1] makes use of the frequency domain version of the image so as to embed watermark information as opposite to spatial domain methods. Discrete cosine transform (DCT) and discrete wavelet transform (DWT) are two of the most frequently used transforms where the frequency domain watermarking algorithms work. In this method an image is transformed from the spatial to frequency domain. Then according to the human perception system, frequency coefficients are arranged into various priorities. Then the magnitude of frequency coefficients is modulated so as to embed watermark bits. The whole process is performed mainly in three steps as explained below.

First step is to perform forward transformation which converts the spatial domain frame into the frequency domain and gives the coefficients. Second step is to modify the frequency coefficients as per the watermark signal. Last step is to perform the inverse transform so as to have modified spatial domain frame.

3.1 Discrete Cosine Transform (DCT)-Based Approach

Discrete cosine transform [2–9] converts frame from the spatial domain into the frequency domain. Eqs. 3.1 and 3.2 show the mathematical representation of forward DCT and reverse DCT, respectively, where $f(x, y)$ is a representation of frame in spatial domain and $F(u, v)$ is a representation in frequency domain. Discrete cosine transform converts a spatial domain 2-D representation into its frequency domain equivalent. There are some of the observations of what appears at the output of DCT. One and the most important observation is that the size of the transformed frequency domain frame is exactly equal to that in the spatial domain frame. Another observation is that the DC coefficient of the DCT transformed matrix is situated at the upper left pixel, and hence the upper left corner of the matrix contains very important part of the information since it contains low-frequency

© Springer International Publishing AG, part of Springer Nature 2019
A. M. Kothari et al., *Watermarking Techniques for Copyright Protection of Videos*, Signals and Communication Technology,
https://doi.org/10.1007/978-3-319-92837-1_3

coefficients. All other pixels other than DC coefficients are called AC coefficients. Another observation is that going toward the right side in zigzag fashion, the frequency increases, and the values of the coefficients start decreasing. Another observation is that the DC coefficient is always an integer, and the range of that would be in between −1024 and 1023, while AC coefficient may be integer or non-integer.

This way of discriminating frequency components makes DCT a powerful tool for the watermarking application. The two-dimensional DCT and inverse DCT are given in Eqs. 3.1 and 3.2, respectively. Figure 3.1 (a) shows the idea of frequency discrimination by DCT.

$$F(u,v) = \alpha(u) \cdot \alpha(v) \sum_{x=0}^{M-1}\sum_{y=0}^{N-1} f(x,y)\cos\left[\frac{(2x+1)u\pi}{2M}\right]\cos\left[\frac{(2y+1)v\pi}{2N}\right] \tag{3.1}$$

where, $\alpha(u) = \sqrt{1/M}$ for $u = 0$; $\alpha(u) = \sqrt{2/M}$ for $u = 1, 2, 3...M - 1$; $\alpha(v) = \sqrt{1/N}$ for $v = 0$; $\alpha(v) = \sqrt{2/N}$ for $v = 1, 2, 3...N - 1$.

$$f(x,y) = \sum_{u=0}^{M-1}\sum_{v=0}^{N-1} \alpha(u) \cdot \alpha(v) \cdot F(u,v)\cos\left[\frac{(2x+1)u\pi}{2M}\right]\cos\left[\frac{(2y+1)v\pi}{2N}\right] \tag{3.2}$$

where $x = 0, 1, 2... M - 1$, $y = 0, 1, 2... N - 1$.

Here F_L denotes low-frequency components, F_M denotes mid-frequency components, and F_H denotes the high-frequency components. Figure 3.2 shows the example application of DCT on an image.

Most important information are situated in the low-frequency components, and when the frame undergoes compression, it is the high-frequency components that are very easily removed, and therefore mid-frequency components are chosen for the watermarking purpose. In the proposed method, two nearby locations having ideal values are chosen from the middle band coefficients for the sake of comparison. From the recommended JPEG quantization table shown in Fig. 3.1b, one can take two locations (5, 2) and (4, 3) which are having identical quantization values.

F_L	F_L	F_L	F_M	F_M	F_M	F_M	F_H
F_L	F_L	F_M	F_M	F_M	F_M	F_H	F_H
F_L	F_M	F_M	F_M	F_M	F_H	F_H	F_H
F_M	F_M	F_M	F_M	F_H	F_H	F_H	F_H
F_M	F_M	F_M	F_H	F_H	F_H	F_H	F_H
F_M	F_M	F_H	F_H	F_H	F_H	F_H	F_H
F_M	F_H	F_H	F_H	F_H	F_H	F_H	F_H
F_H	F_H	F_H	F_H	F_H	F_H	F_H	F_H

16	11	10	26	24	40	51	61
12	12	14	19	26	58	60	55
14	13	16	24	40	57	69	56
14	17	22	29	51	87	80	62
18	22	37	56	68	109	103	77
24	35	55	64	81	104	113	92
49	64	78	87	103	121	120	101
72	92	95	98	112	100	103	99

(a)   (b)

Fig. 3.1 (a) Frequency discrimination by DCT. (b) Quantization values used in JPEG compression scheme

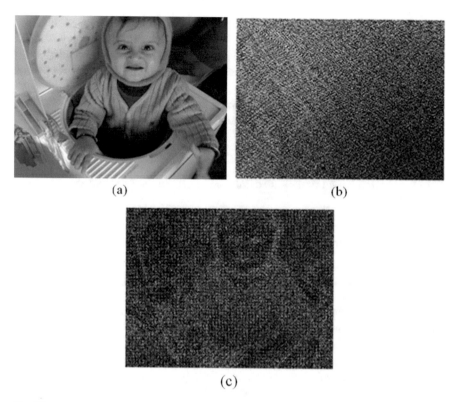

Fig. 3.2 Energy concentration of DCT. (**a**) Original image. (**b**) DCT of the image. (**c**) Blockwise DCT of image

The advantage of taking these two locations is that if any of the location values is scaled, then the other is also scaled by the same factor and hence preserves their relative size.

3.1.1 Watermark Embedding Process

The following are the sequence of events that takes place when the watermark is embedded in this method.

Step 1.　Original video is broken into a number of frames.

Step 1.　Colorspace conversion is done from RGB to YCbCr.

Step 2.　Y frame is divided into non-overlapping blocks.

Step 3.　DCT is applied to each and every block.

Step 4.　Two diagonal components, i.e., (5, 2) and (4, 3), from the mid-frequency band are modified according to the property of the watermark. If watermark is black, it is taken care that (5, 2) is greater than (4, 3), and if it is not, they are swapped. Similarly if the watermark is white, it is taken care that (5, 2) is less than (4, 3), and if it is not, they are swapped. It is also taken care that the difference between the two mid-frequency bands is kept more than the gain factor, and if it is not found to be greater, then the gain factor is added and subtracted so as to maintain the difference greater than gain factor.

Step 5.   Inverse DCT is applied to each modified block.

Step 6.   Inverse colorspace conversion is applied so as to get modified RGB frame.

Step 7.   Steps 2 to 7 are executed for the next frame, and the process continues until the last frame.

Step 8.   All modified frames are combined to get the modified video.

3.1.2   Watermark Extraction Process

The steps described below show the methodology of extracting the watermark back from the watermarked video.

Step 1.   Watermarked video is broken into a number of frames.

Step 2.   Colorspace conversion is done from RGB to YCbCr.

Step 3.   Y frame is divided into non-overlapping blocks.

Step 4.   DCT is applied to each and every block.

Step 5.   Two diagonal components, i.e., (5, 2) and (4, 3), from the mid-frequency band are checked and compared. If (5, 2) is found to be greater than (4, 3), the assigned watermark bit is black, and if (5, 2) is less than (4, 3), the assigned watermark bit is white. Hence the watermark is recovered.

Step 6.   Steps 2 to 5 are executed for the next frame, and the process continues until the last frame so as to get all watermarks.

3.1.3   Results of Imperceptibility Test

For imperceptibility test of DCT-based approach, the cover digital video with a size of 512 × 512 pixels and monochrome messages with a size of 64 × 64 pixels are taken. Here, video frame is divided into 8 × 8 non-overlapping blocks. The one bit of the message is embedded into each block of video frame to the generated watermarked frame using gain factor. Figure 3.3 shows the results of DCT-based approach with gain factor 100 and block size 8. The quality measures for DCT-based approach for gain factor 100 and block size be 8 are summarized in Table 3.1.

In the watermark embedding process, the gain factor is playing an important role. The visual quality of watermarked video and extracted messages is depending on the value of the gain factor. Figure 3.4 shows results of the method on Frame 1 considering various values of the gain factor. Table 3.2 shows the results of watermarking Frame 1 with various gain factors.

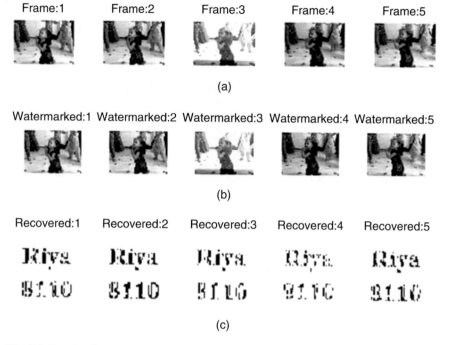

Fig. 3.3 Results of imperceptibility test of DC-based approach. (a) Original video frames, (b) Watermarked video frames, (c) Extracted messages

Table 3.1 Quality measure values for DCT-based approach

Video frame	PSNR (dB)	Correlation
Frame 1	34.8816	0.8996
Frame 2	34.9053	0.9248
Frame 3	33.6815	0.8002
Frame 4	35.1199	0.8562
Frame 5	34.863	0.9033

3.1.4 Results of Robustness Test

For robustness test, various types of watermarking attacks are applied on watermarked video frames, and messages are extracted from these corrupted watermarked video frames. If message is successfully extracted from the corrupted frames, then approach is said to be robust. Here various types of watermarking attacks such as filtering attacks, addition of noise, geometric attacks, and other signal processing with different values are applied on watermarked video frame.

Figures 3.5, 3.6, 3.7, and 3.8 show the results of robustness test for DCT-based approach against different filters such as average filter, Gaussian low-pass filter median filter, and high-pass filter, respectively. Here the numbers on the upper side of the video frames show PSNR values and that in the lower side of the frames show variants of particular attacks. Similarly the numbers on the upper side of the recovered watermark show correlation values.

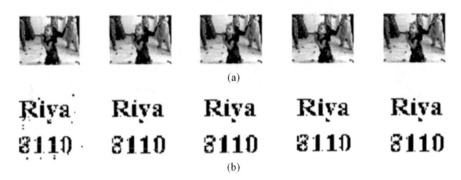

(a)

(b)

Fig. 3.4 Results of DCT-based approach with various gain factor values like 10, 30, 50, 70, and 90. (**a**) Watermark video Frame 1, (**b**) Extracted messages

Table 3.2 Results of DCT-based approach on Frame 1 using various gain factors

Gain factor (k)	PSNR (dB)	Correlation
10	46.0311	0.8705
30	42.8575	0.8996
50	39.7859	0.8996
70	37.5468	0.8996
90	35.6277	0.8996
100	34.8816	0.8996

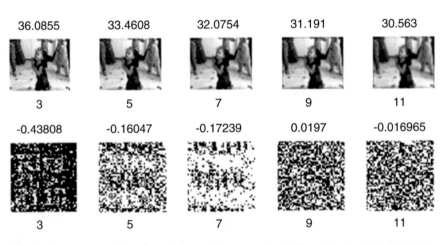

Fig. 3.5 Watermarked video frame and extracted message for DCT-based approach under average filtering attack with various mask sizes

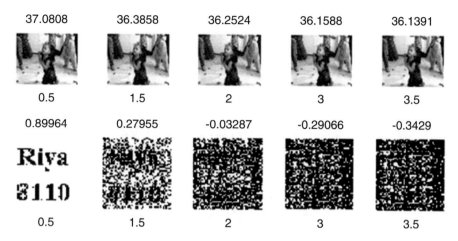

Fig. 3.6 Watermarked video frame and extracted message for DCT-based approach under Gaussian low-pass filtering attack with various standard deviations

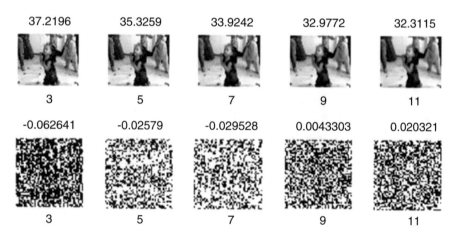

Fig. 3.7 Watermarked video frame and extracted message for DCT-based approach under median filtering attack with various mask sizes

Fig. 3.8 Watermarked video frame and extracted message for DCT-based approach under high-pass filtering attack

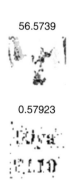

The compression is very basic attacks which affect any multimedia data when it is transmitted over communication channel. Figure 3.9 shows the results of robustness test for DCT-based approach against JPEG compression with various quality factors. Figures 3.10, 3.11, and 3.12 show the results of robustness test for DCT-based approach against attacks such as color reduction, histogram equalization, and motion blurring.

Figures 3.13, 3.14, and 3.15 show the results of robustness test for DCT-based approach against various noise addition attacks such as Gaussian noise, salt and pepper noise, and speckle noise. Figures 3.16 and 3.17 show the results of robustness test for DCT-based approach against geometric attacks such as rotation and cropping.

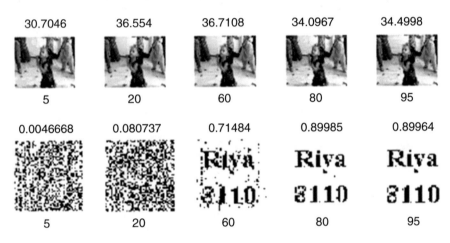

Fig. 3.9 Watermarked video frame and extracted message for DCT-based approach under compression attack with various quality values

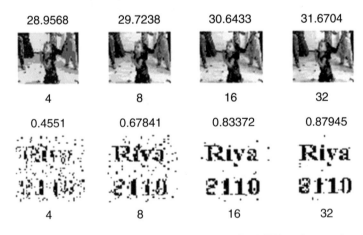

Fig. 3.10 Watermarked video frame and extracted message for DCT-based approach under color reduction attack with various number of colors

Fig. 3.11 Watermarked video frame and extracted message for DCT-based approach under histogram equalization attack

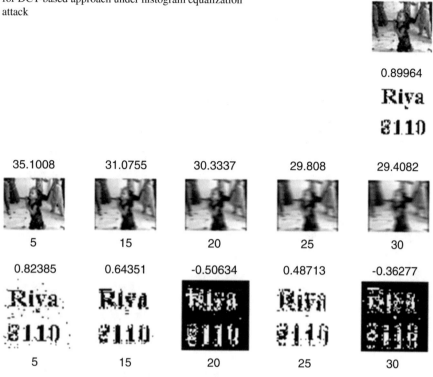

Fig. 3.12 Watermarked video frame and extracted message for DCT-based approach under motion blurring attack

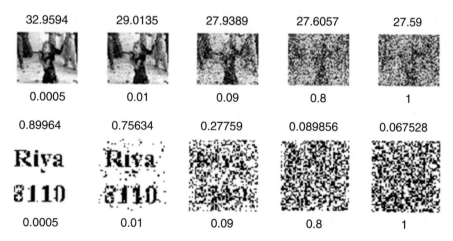

Fig. 3.13 Watermarked video frame and extracted message for DCT-based approach under Gaussian noise attack with zero mean and various variances

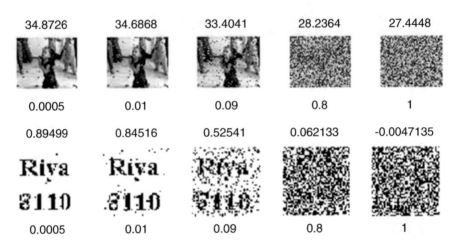

Fig. 3.14 Watermarked video frame and extracted message for DCT-based approach under salt and pepper noise attack with various variances

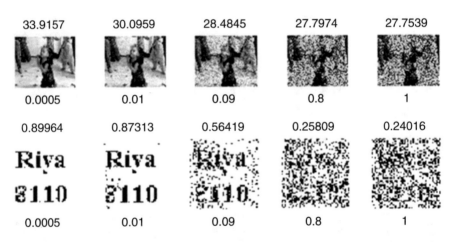

Fig. 3.15 Watermarked video frame and extracted message for DCT-based approach under speckle noise attack with various variances

3.1.5 Observation on Obtain Results

The following are some of the observations made after successfully implementing both embedding and extracting algorithm. Here the gain factor of 100 is assumed for the sake of observations in terms of perceptibility and robustness. The higher is the value of PSNR, the higher is the perceptibility, and the higher is the value of correlation, the higher is the robustness.

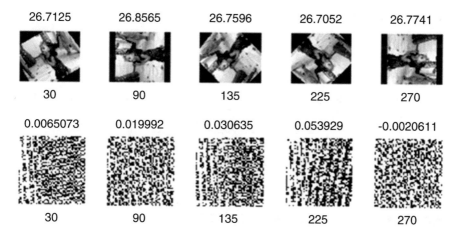

Fig. 3.16 Watermarked video frame and extracted message for DCT-based approach under rotation attack with various angles

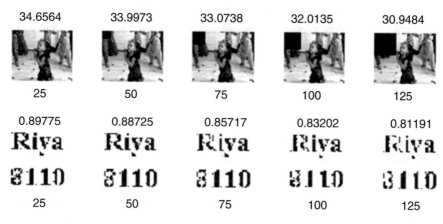

Fig. 3.17 Watermarked video frame and extracted message for DCT-based approach under cropping attack with various crop regions

- Perceptibility in this method decreases with the increase in gain factor.
- Robustness increases with the increase in gain factor.
- The frames look visibly fine if the resultant PSNR is above 28 dB. The message seems visibly identifiable if the resultant correlation is greater than 0.50.
- This method is not robust against average filtering, median filtering, and rotation attacks.
- This method is partially robust against Gaussian low-pass filtering, compression, Gaussian noise, salt and pepper noise, and speckle noise attacks.
- This method is fully robust against color reduction, histogram equalization, motion blurring, cropping, and high-pass filtering attacks.

3.1.6   Comparison with Correlation-Based Approach

The comparison of DCT-based approach with correlation-based approach is given in this section.

- At same gain factor, perceptibility difference is very high with very small difference in robustness.
- Less robustness is achieved in compression and speckle noise attacks.
- More robustness is achieved in Gaussian low-pass filtering, linear motion of camera, and high-pass filtering attacks.

3.2   Discrete Wavelet Transform (DWT)-Based Approach

In most of the applications that includes the field of image processing, the wavelet transform has very important contributions to make the application smooth and fruitful. Compression, signal analysis, digital watermarking, and signal processing have been some of the applications made practical in this field of study in the past few decades.

Waves are periodic in nature and are oscillating with respect to time or space. On the other side, wavelets are localized waves with its energy concentrated in time or space, and they are used for the purpose of analyzing a signal. Actually the discrete wavelet transform does the convolution operation between the 1-D and 2-D signals with particular instances of wavelets at various time scales and positions. The DWT [8, 10–17] is based on sub-band coding, is easy to implement, does require limited time and resources, and yields fast computations of wavelet transform. The DWT is a combined process of filtering and subsampling where subsampling may be upsampling or downsampling. Filtering operation determines the resolution, known to be a measurement about how much information the signal does contain, while subsampling operation determines the scale of the signal. A series of successive low-pass and high-pass filters are used to compute multilevel DWT. At each decomposition level, frequency resolution is doubled as the uncertainty in frequency is reduced by half and time resolution is made half means if the signal has originally of 500 samples, it reduces to 250 samples at the end of first decomposition level. One of the beautiful observations is that at high frequency, we have very good time resolution, whereas at low frequencies frequency resolution is observed good.

To understand the basic idea of the DWT, let us focus on one-dimensional signal. The signal is passed through a low-pass filter and a high-pass filter so as to get both high- and low-frequency parts of the signal. High-frequency part contains edge components wherein low-frequency part contains information components. The same process is repeated for the low-frequency part so as to get the second level low- and high-frequency components. The number of decomposition levels depends on the application of interest. So far as watermarking and compression are concerned, a maximum of five levels of decomposition are computed. The original signal can also be reconstructed from the knowledge of DWT coefficients. This process of reconstruction (synthesis) is called the inverse DWT (IDWT).

So goes a popular saying, any signal contains its most important and informative part in its low-frequency component and that is the reason why low-frequency components are very important. On the other hand, the high-frequency components are of less importance. Consider the human voice. If high-frequency components are removed from a song, it would sound different, but one can still identify the saying. However, if low-frequency components are removed, one would be able to hear garbage only.

In wavelet analysis two words are frequent, i.e., *approximations* and *details*. The approximations are the high-scale, low-frequency components of the signal. The details are the low-scale, high-frequency components. Figure 3.18 shows the first stage of the decomposition wherein signal is applied to low-pass and high-pass filters.

If the original signal is of size 1x1000, then the size of each of the approximation and detail component would be 1 × 1000. So the output contains twice the samples compared to the input. So the output of both of the filters is downsampled by two so that each of the output would have half the size of the original signal and hence the total size equals to that of the original signal. Figure 3.19 shows the concept. The decomposition or analysis process with downsampling produces DWT coefficients.

Mathematically, the DWT and IDWT can be stated as follows. Eqs. 3.3 and 3.4 are equations of low- and high-pass filters, respectively.

$$H(w) = \sum_k hk \cdot e^{-jkw}$$

(3.3)

$$G(w) = \sum_k gk \cdot e^{-jkw}$$

(3.4)

The successive approximation components can be iteratively decomposed so that one signal can be divided into many components having lower resolution. This is said to be the wavelet decomposition tree, and it is shown in Fig. 3.20.

Fig. 3.18 Filtering or decomposition process at its most basic level

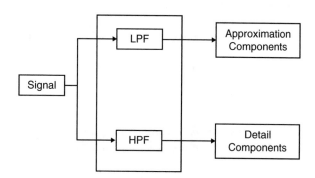

Fig. 3.19 Analysis with
downsampling

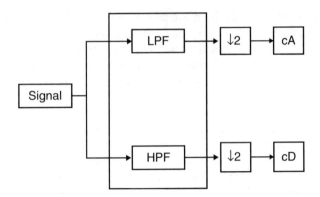

For a two-dimensional image $F(x, y)$, the forward and reverse decomposition can be done by applying the DWT and IDWT first on dimension x, and then the same process can be performed for the other dimension y. This results in the representation of the image which is pyramidal in nature. This kind of 2-D DWT decomposes the image into four parts, namely, approximation component, horizontal detail component, vertical detail components, and diagonal detail components. Figure 3.10 describes the basic decomposition steps for images (Fig. 3.21).

An example of the two-level DWT is demonstrated in Fig. 3.22.

Here concepts of DWT and the correlation are combined so as to embed the watermark into the video.

3.2.1   Watermark Embedding Process

Steps for the video watermarking using DWT are explained herewith:

Step 1   Original video is broken into a number of frames.
Step 2   Two random sequences are designed and named pn_sequence_one and pn_sequence_zero.
Step 3   Colorspace conversion is done from RGB to YCbCr.
Step 4   Discrete wavelet transform is applied to Y frame, and HL component is chosen for the embedding purpose.
Step 5   HL component is divided into non-overlapping blocks.
Step 6   If the message bit is zero, watermark block is filled with pn_sequence_zero; otherwise it is filled with pn_sequence_one. Same thing is repeated for all blocks.
Step 7   Original HL component is added with the weighted watermark block wherein the weight is called the gain factor.
Step 8   Inverse DWT is applied to get watermarked Y frame.
Step 9   Inverse colorspace conversion is applied so as to get the modified RGB frame.
Step 10   Steps 3 to 9 are executed for the next frame, and the process continues until the last frame.

Watermarked video is obtained by combining all watermarked frames.

Fig. 3.20 Multiple-level decomposition or analysis

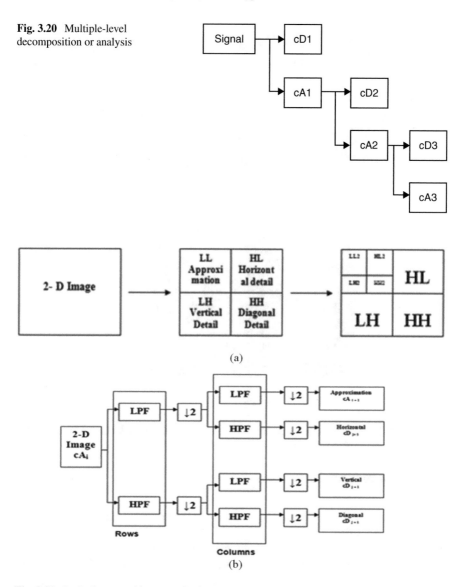

(a)

(b)

Fig. 3.21 Basic decomposition steps for images

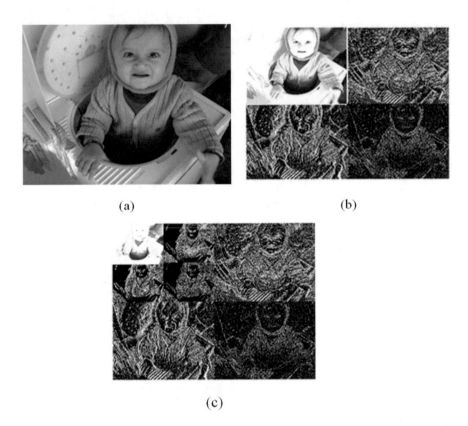

(a)   (b)

(c)

Fig. 3.22 Decomposition of an image using DWT. (**a**) Original image, (**b**) first-level decomposition, (**c**) second-level decomposition

3.2.2   Watermark Extraction Process

The steps described below show the methodology of extracting the watermark back from the watermarked video.

Step 1.   Watermarked video is broken into a number of frames.
Step 2.   Two random sequences are designed and named pn_sequence_one and pn_sequence_zero. These sequences must be the same as that at the embedding side.
Step 3.   Colorspace conversion is done from RGB to YCbCr.
Step 4.   Discrete wavelet transform is applied to Y frame, and HL component is chosen for the extraction purpose.
Step 5.   HL component is divided into non-overlapping blocks.
Step 6.   Correlation of a block with both the sequences is calculated. If correlation is found more in case of pn_sequence_zero, then message bit is assigned 0; otherwise it is assigned 1. The same operation is done with all blocks and watermark is extracted.
Step 7.   Steps 3 to 6 are executed for the next frame, and the process continues until the last frame.

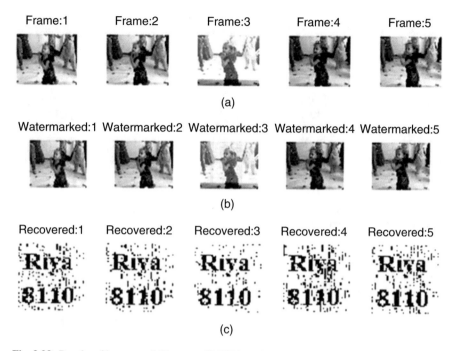

Fig. 3.23 Results of imperceptibility test of DWT-based approach. (a) Original video frames, (b) Watermarked video frames, (c) Extracted messages

Table 3.3 Quality measure values for DWT-based approach

Video frame	PSNR (dB)	Correlation
Frame 1	34.603	0.7359
Frame 2	35.1536	0.7527
Frame 3	28.9982	0.8119
Frame 4	37.7939	0.6834
Frame 5	35.313	0.7221

3.2.3 Results of Imperceptibility Test

For imperceptibility test of DWT-based approach, the cover digital video with a size of 512 × 512 pixels and monochrome messages with a size of 64 × 64 pixels are taken. Here, video frame is divided into 8 × 8 non-overlapping blocks. The one bit of message is embedded into each block of video frame to the generated watermarked frame using gain factor. Figure 3.23 shows the results of DWT-based approach with gain factor 100 and block size 8. The quality measures for DCT-based approach for gain factor 100 and block size be 8 are summarized in Table 3.3.

In the watermark embedding process, the gain factor is playing an important role. The visual quality of watermarked video and extracted messages is depending

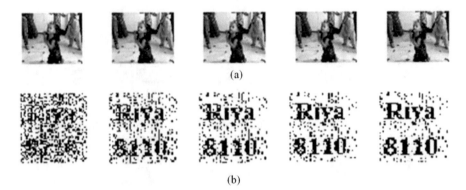

(a)

(b)

Fig. 3.24 Results of DWT-based approach with various gain factor values like 10, 30, 50, 70, and 90. (**a**) Watermark video Frame 1, (**b**) Extracted messages

Table 3.4 Results of DWT-based approach on Frame 1 using various gain factors

Gain factor (k)	PSNR (dB)	Correlation
10	46.9929	0.2991
30	44.3318	0.5391
50	41.0474	0.6324
70	38.0773	0.6845
90	35.6396	0.7248
100	34.603	0.7359

on the value of the gain factor. Figure 3.24 shows results of the method on Frame 1 considering various values of the gain factor. Table 3.4 shows the results of watermarking Frame 1 with various gain factors.

3.2.4   Results of Robustness Test

For robustness test, various types of watermarking attacks are applied on watermarked video frames, and messages are extracted from these corrupted watermarked video frames. If the message is successfully extracted from the corrupted frames, then the approach is said to be robust. Here various types of watermarking attacks such as filtering attacks, addition of noise, geometric attacks, and other signal processing with different values are applied on watermarked video frame.

Figures 3.25, 3.26, 3.27, and 3.28 show the results of robustness test for DWT-based approach against different filters such as average filter, Gaussian low-pass filter median filter, and high-pass filter, respectively. Here the numbers on the upper side of the video frames show PSNR values and that in the lower side of the frames show variants of particular attacks. Similarly the numbers on the upper side of the recovered watermark show correlation values.

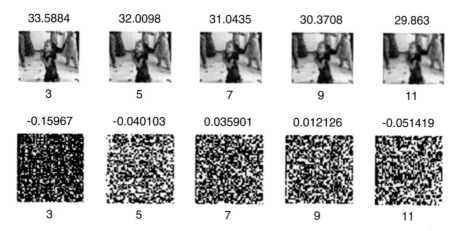

Fig. 3.25 Watermarked video frame and extracted message for DWT-based approach under average filtering attack with various mask sizes

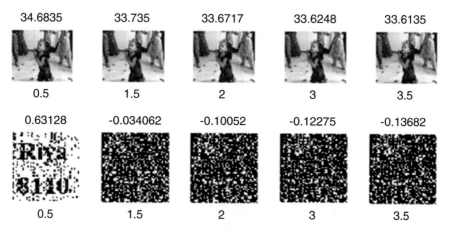

Fig. 3.26 Watermarked video frame and extracted message for DWT-based approach under Gaussian low-pass filtering attack with various standard deviations

The compression is very basic attacks which affects any multimedia data when it is transmitted over the communication channel. Figure 3.29 shows the results of robustness test for DWT-based approach against JPEG compression with various quality factors. Figures 3.30, 3.31, and 3.32 show the results of robustness test for DWT-based approach against attacks such as color reduction, histogram equalization, and motion blurring.

Figures 3.33, 3.34, and 3.35 show the results of robustness test for DWT-based approach against various noise addition attacks such as Gaussian noise, salt and pepper noise, and speckle noise. Figures 3.36 and 3.37 show the results of robustness test for DWT-based approach against geometric attacks such as rotation and cropping.

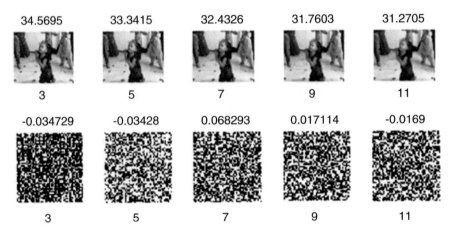

Fig. 3.27 Watermarked video frame and extracted message for DWT-based approach under median filtering attack with various mask sizes

Fig. 3.28 Watermarked video frame and extracted message for DWT-based approach under high-pass filtering attack

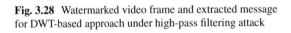

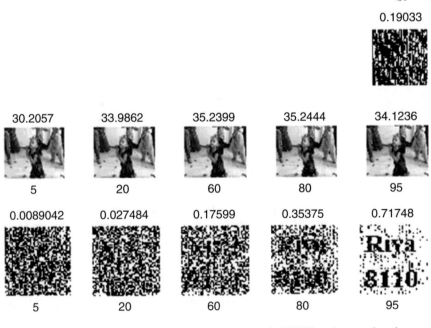

Fig. 3.29 Watermarked video frame and extracted message for DWT-based approach under compression attack with various quality values

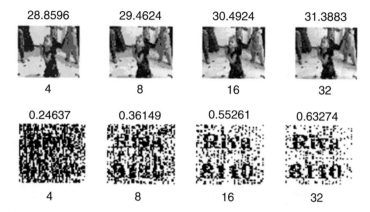

Fig. 3.30 Watermarked video frame and extracted message for DWT-based approach under color reduction attack with various number of colors

Fig. 3.31 Watermarked video frame and extracted message for DWT-based approach under histogram equalization attack

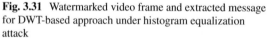

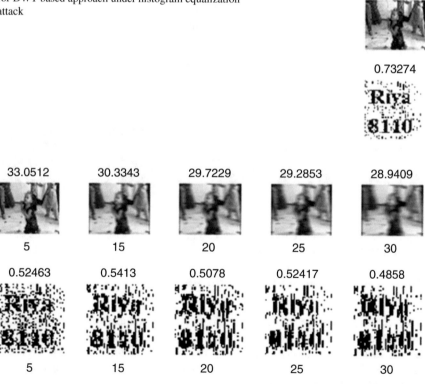

Fig. 3.32 Watermarked video frame and extracted message for DWT-based approach under motion blurring attack

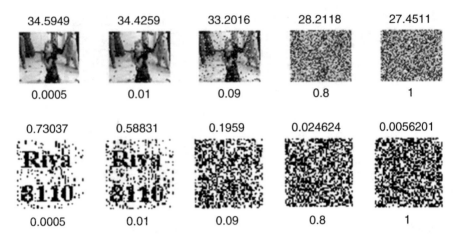

Fig. 3.33 Watermarked video frame and extracted message for DWT-based approach under Gaussian noise attack with zero mean and various variances

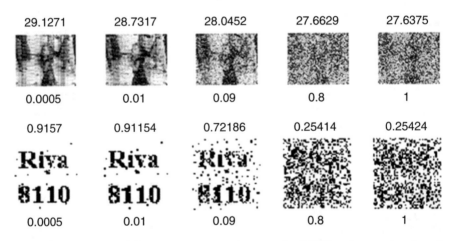

Fig. 3.34 Watermarked video frame and extracted message for DWT-based approach under salt and pepper noise attack with various variances

3.2.5   Observation on Obtain Results

The following are some of the observations made after successfully implementing both embedding and extracting algorithm. Here the gain factor of 100 is assumed for the sake of observations in terms of perceptibility and robustness. The higher is the value of PSNR, the higher is the perceptibility, and the higher is the value of correlation, the higher is the robustness.

- Perceptibility in this method decreases with the increase in gain factor.
- Robustness increases with the increase in gain factor.

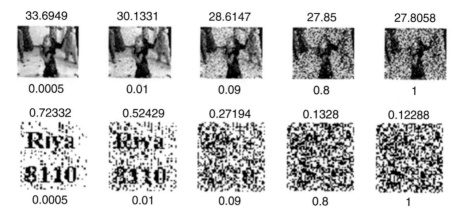

Fig. 3.35 Watermarked video frame and extracted message for DWT-based approach under speckle noise attack with various variances

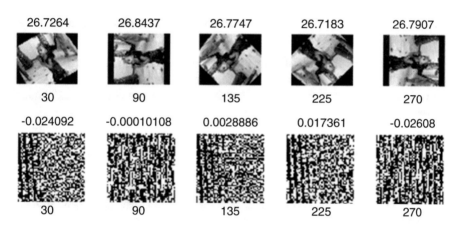

Fig. 3.36 Watermarked video frame and extracted message for DWT-based approach under rotation attack with various angles

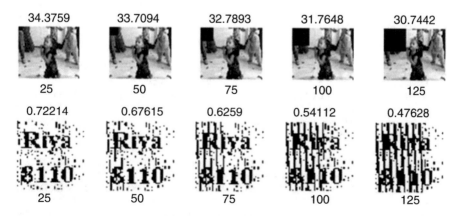

Fig. 3.37 Watermarked video frame and extracted message for DWT-based approach under cropping attack with various crop regions

- The frames look visibly fine if the resultant PSNR is above 28 dB. The message seems visibly identifiable if the resultant correlation is greater than 0.50.
- This method is not robust against average filtering, median filtering, rotation, and high-pass filtering attacks.
- This method is partially robust against Gaussian low-pass filtering, compression, color reduction, Gaussian noise, salt and pepper noise, and speckle noise attacks.
- This method is fully robust against histogram equalization, linear motion of camera, and cropping attacks.

3.2.6 Comparison with Correlation and DCT-Based Approaches

The comparison of DWT-based approach with correlation-based approach and DCT-based approach is given in this section.

- Out of the three, the perceptibility is highest in this method at the same gain factor with a noticeable difference in the robustness.
- Robustness is less in this method in almost all attacks as compare to the other two methods.

Bibliography

1. D. He, Q. Sun and T. Qi, A semi-fragile object based video authentication system. Proc. Int. Symp. Circuits Syst. 814–817 (2003)
2. S. Arena, M. Caramma, R. Lancini, Digital watermarking applied to MPEG-2 coded video sequences exploiting space and frequency masking. Proc. Int. Conf. Image Proces. 2, 796–799 (2000)
3. J.R. Hernandez, M. Amado, F. Perez-Gonzalez, DCT-domain watermarking techniques for still image: Detector performance analysis and a new structure. IEEE Trans. Image Proces. 9, 55–68 (2000)
4. L. Chun-Shien, H.-Y.M. Liao, Video object-based watermarking: A rotation and flipping resilient scheme. Proc. Int. Conf. Image Proces. (2001)
5. R.O. Preda, D.N. Vizireanu, Blind watermarking capacity analysis of MPEG2 coded video. Proc. Conf. Telecommun. Modern Satell. Cable Broadcast. Services, Serbia. 465–468 (2007)
6. A. Alper Koz, A. Alatan, Oblivious spatio-temporal watermarking of digital video by exploiting the human visual system. IEEE Trans. Circuits Syst. Video Technol. 18(3), 326–337 (2008)
7. T. Sridevi, B. Krishnaveni, V. Vijaya Kumar, Y. Rama Devi, A video watermarking algorithm for MPEG videos. A2CWiC 2010 - Amrita ACM-W Celebration Women Comput. 16–17 Sept 2010
8. Y. Ding, X. Zheng, Y. Zhao, G. Liu, A video watermarking algorithm resistant to copy attack. Proc. Third Int. Symp. Electron. Commerce Secur. 29 July 2010
9. F. Huang, Z.-H. Guan, A hybrid SVD-DCT watermarking method based on LPSNR. Pattern Recogn. Lett. 25, 1769–1775 (2004)
10. M. ejima, A. Miyazaki, A wavelet-based watermarking for digital images and video. IEEE Int Conf Image Proces. 678–681 (2000)

11. C.V. Serdean, M.A. Ambroze, M. Tomlinson and G. Wade, Combating geometrical attacks in a dwt based blind video watermarking system. IEEE Region 8 Int. Symp. Video/Image Proces. Multimed Commun. Zadar, Croatia. 263–266, 16–19 June 2002

12. F. Liang, F. Yanmei, A DWT-Based Video Watermarking Algorithm Applying DS-CDMA. IEEE Region 10 Conf. TENCON 2006. 14–17 Nov 2006

13. E. Ersin, Robust mpeg video watermarking in wavelet domain. Trakya Univ. J. Sci. **8**(2), 87–93 (2007)

14. A. Essaouabi, E. Ibnelhaj, A 3D Wavelet -Based Method for Digital Video Watermarking. Proc. 4th IEEE Int. Inf. Hiding Multimed. Signal Proces., 29–31 July 2009

15. K. Raghavendra, K.R. Chetan, A blind and robust watermarking scheme with scrambled watermark for video authentication. Proc IEEE Int. Conf. Int. Multimed. Services Archit. Appl. 9–11 Dec 2009

16. J. Hussein, A. Mohammed, Robust Video Watermarking using Multi-Band Wavelet Transform. IJCSI Int. J. Comput. Sci. Issues. **6**(1), 44–49 (2009)

17. S.A.K. Mostafa, A.S. Tolba, F.M. Abdelkader, H.M. Elhindy, Video watermarking scheme based on principal component analysis and wavelet transform. IJCSNS Int. J. Comput. Sci. Netw. Secur. **9**(8), 45–52 (2009)

Chapter 4
Singular Value Decomposition (SVD)-Based Video Watermarking

Singular value decomposition (SVD) [1–7] is a numerical technique based on the linear algebra, and it is used to diagonalize matrices in numerical analysis. There are lots of areas where SVD finds its application. When SVD is applied to image A of size M × N, three matrices are found, namely, U, V, and S, whose properties are:

- It can be represented as $A = USV^T$.
- U and V matrices are called unitary matrices having sizes M × M and N × N, respectively.
- S matrix is called diagonal matrix having size M × N.
- U and V matrix are labeled left and right singular values of the matrix A, respectively.
- S matrix, singular matrix, is very important for watermarking purpose, and entries in this matrix are arranged diagonally and in ascending order.
- One of the most important properties of the singular values is that they are very much stable, and hence if small change is made in the value of the cover medium, its singular values do not have any significant change.
- SVD efficiently represents the basic properties of an image which are being algebraic in nature. Here the brightness of the image is given by singular values, and the geometric characteristic of the image is represented by singular vectors.
- An image matrix has many small singular values compared with the first singular value. Even if these singular values are ignored, there is no perceptibly significant change in the reconstructed image.

One example of the SVD process is explained in Fig. 4.1.

The main advantage of using SVD in digital watermarking is that one can embed binary as well as grayscale watermark behind the video which was not at all possible with the previous methods.

© Springer International Publishing AG, part of Springer Nature 2019
A. M. Kothari et al., *Watermarking Techniques for Copyright Protection of Videos*, Signals and Communication Technology,
https://doi.org/10.1007/978-3-319-92837-1_4

Original Matrix		
1	2	3
4	5	6
7	8	9
10	11	12

U Matrix			
0.140877	-0.82471	-0.54556	0.048576
0.343946	-0.42626	0.691212	-0.47141
0.547016	-0.02781	0.254268	0.797087
0.750086	0.370637	-0.39992	-0.37426

V Matrix		
0.504533	0.760776	0.408248
0.574516	0.057141	-0.8165
0.644498	-0.64649	0.408248

S Matrix		
25.46241	0	0
0	1.290662	0
0	0	1.46E-15
0	0	0

Fig. 4.1 Example of SVD

4.1 SVD-Based Approach with Binary Message

In this method binary message same as that used in all methods in Chaps. 2 and 3 is being embedded in the video.

4.1.1 Watermark Embedding Process

The following is the sequence of events that take place when binary message is embedded behind the video using SVD as a main tool.

Step 1.   Original video is broken into a number of frames.

Step 2.   A frame is taken, and colorspace conversion is applied to convert RGB frame into YCbCr frame.

Step 3.   Y frame is selected for the embedding purpose.

Step 4.   SVD is applied on the selected Y frame.

Step 5.   Watermark is rescaled to the size of the singular component, i.e., S.

Step 6.   Singular component is modified as $S = S + K * W$ where W is the watermark and K is the gain factor.

Step 7.   Again SVD is applied on the modified singular component.

Step 8.   Selected sub-band is modified as New_Value = $U*Modified\_S*V^T$.

Step 9.   Inverse colorspace conversion is applied.

Step 10. Steps 2 to 9 are executed until the end of all frames.

Step 11. All watermarked frames are combined to have watermarked video.

4.1.2 Watermark Extraction Process

The following is the stepwise representation of extracting the message at the receiver end from the video.

Step 1.   Watermarked video is broken into a number of frames.
Step 2.   A frame is taken, and colorspace conversion is applied to convert RGB frame into YCbCr frame.
Step 3.   Y frame is selected for the extracting purpose.
Step 4.   SVD is applied on the selected Y frame.
Step 5.   Singular part is resized to have the size same as the message so as to have $D = U*S*V^T$.
Step 6.   Watermark is generated by applying (D - S) / K.
Step 7.   The process is repeated for every frame and all watermarks are retrieved.

4.1.3   Results of Imperceptibility Test

For imperceptibility test of SVD-based approach for binary message, the cover digital video with a size of 512×512 pixels and monochrome messages with a size of 512×512 pixels are taken. The one bit of message is embedded into singular values of video frame to generated watermarked frame using gain factor. Figure 4.2 shows the results of SVD-based approach for binary message with gain factor 100. The

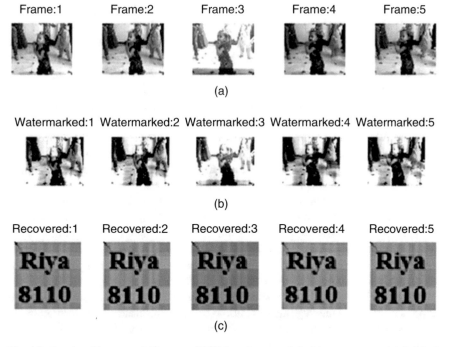

Fig. 4.2 Results of imperceptibility test of SVD-based approach for binary message. (**a**) Original video frames, (**b**) Watermarked video frames, (**c**) Extracted messages

Table 4.1 Quality measure values for SVD-based approach for binary message

Video frame	PSNR (dB)	Correlation
Frame 1	28.8693	0.9779
Frame 2	28.8484	0.9720
Frame 3	28.8883	0.9776
Frame 4	29.0698	0.9731
Frame 5	28.6509	0.9764

quality measures for SVD-based approach for binary message for gain factor 100 are summarized in Table 4.1.

In the watermark embedding process, the gain factor is playing an important role. The visual quality of watermarked video and extracted messages is depending on the value of gain factor. Figure 4.3 shows results of the method on Frame 1 considering various values of the gain factor. Table 4.2 shows the results of watermarking Frame 1 with various gain factors.

4.1.4   Results of Robustness Test

For robustness test, various types of watermarking attacks are applied on watermarked video frames, and messages are extracted from these corrupted watermarked video frames. If the message is successfully extracted from the corrupted frames, then the approach is said to be robust. Here various types of watermarking attacks such as filtering attacks, addition of noise, geometric attacks, and other signal processing with different values are applied on watermarked video frame.

Figures 4.4, 4.5, 4.6, and 4.7 show the results of robustness test for SVD-based approach for binary message against different filters such as average filter, Gaussian low-pass filter median filter, and high-pass filter, respectively. Here the numbers on the upper side of the video frames show PSNR values and that in the lower side of the frames show variants of particular attacks. Similarly the numbers on the upper side of the recovered watermark show correlation values.

The compression is very basic attacks which affect any multimedia data when it is transmitted over communication channel. Figure 4.8 shows the results of robustness test for SVD-based approach for binary message against JPEG compression with various quality factors. Figures 4.9, 4.10, and 4.11 show the results of robustness test for SVD-based approach for binary message against attacks such as color reduction, histogram equalization, and motion blurring.

Figures 4.12, 4.13, and 4.14 show the results of robustness test for SVD-based approach for binary message against various noise addition attacks such as Gaussian noise, salt and pepper noise, and speckle noise. Figures 4.15 and 4.16 show the results of robustness test for SVD-based approach for binary message against geometric attacks such as rotation and cropping.

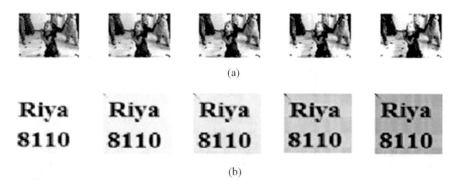

(a)

(b)

Fig. 4.3 Results of SVD-based approach for binary message with various gain factor values like 10, 30, 50, 70, and 90. (**a**) Watermark video Frame 1, (**b**) Extracted messages

Table 4.2 Results of SVD-based approach for binary message on Frame 1 using various gain factors

Gain factor (k)	PSNR (dB)	Correlation
10	44.9324	0.9913
30	32.4366	0.9906
50	30.1491	0.9894
70	29.2756	0.9877
90	28.9323	0.9819
100	28.8693	0.9779

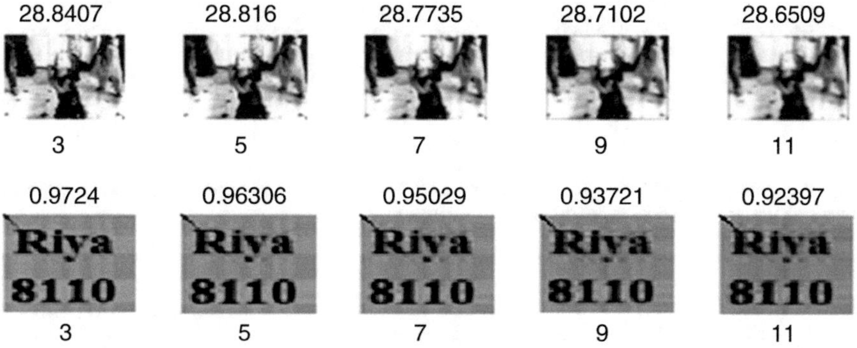

28.8407	28.816	28.7735	28.7102	28.6509
3	5	7	9	11

0.9724	0.96306	0.95029	0.93721	0.92397
3	5	7	9	11

Fig. 4.4 Watermarked video frame and extracted message for SV-based approach for binary message under average filtering attack with various mask sizes

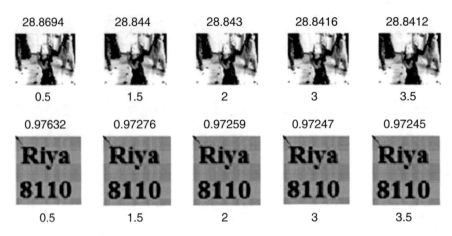

Fig. 4.5 Watermarked video frame and extracted message for SVD-based approach for binary message under Gaussian low-pass filtering attack with various standard deviations

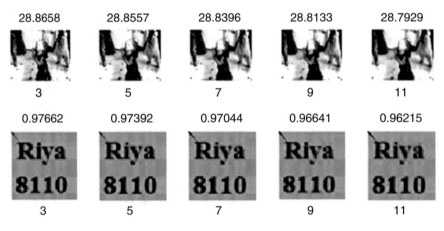

Fig. 4.6 Watermarked video frame and extracted message for SVD-based approach for binary message under median filtering attack with various mask sizes

Fig. 4.7 Watermarked video frame and extracted message for SVD-based approach for binary message under high-pass filtering attack

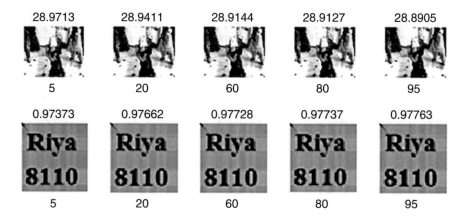

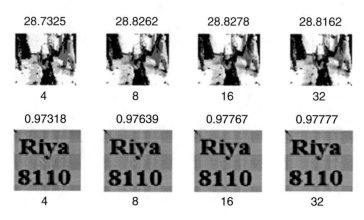

Fig. 4.8 Watermarked video frame and extracted message for SVD-based approach for binary message under compression attack with various quality values

Fig. 4.9 Watermarked video frame and extracted message for SVD-based approach for binary message under color reduction attack with various number of colors

Fig. 4.10 Watermarked video frame and extracted message for SVD-based approach for binary message under histogram equalization attack

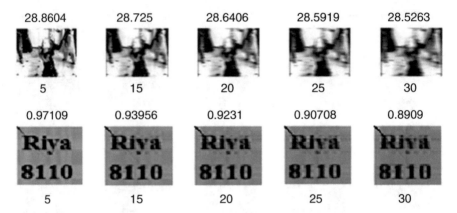

Fig. 4.11 Watermarked video frame and extracted message for SVD-based approach for binary message under motion blurring attack

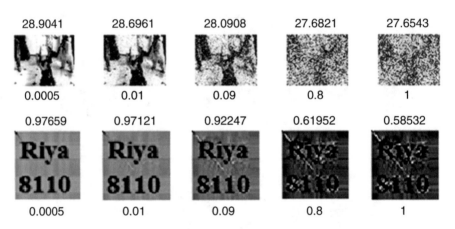

Fig. 4.12 Watermarked video frame and extracted message for SVD-based approach for binary message under Gaussian noise attack with zero mean and various variances

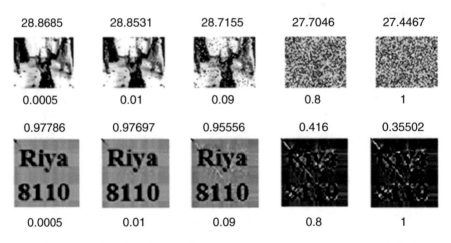

Fig. 4.13 Watermarked video frame and extracted message for SVD-based approach for binary message under salt and pepper noise attack with various variances

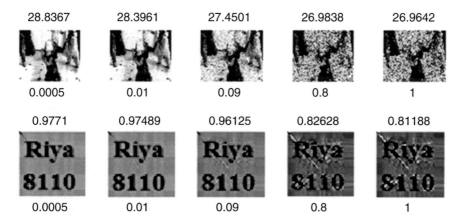

Fig. 4.14 Watermarked video frame and extracted message for SVD-based approach for binary message under speckle noise attack with various variances

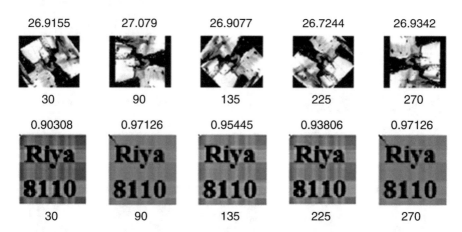

Fig. 4.15 Watermarked video frame and extracted message for SVD-based approach for binary message under rotation attack with various angles

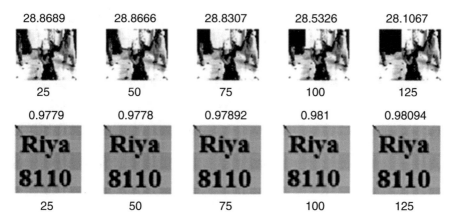

Fig. 4.16 Watermarked video frame and extracted message for SVD-based approach for binary message under cropping attack with various crop regions

4.2   SVD-Based Approach with Grayscale Message

In this method, a grayscale message as shown in Fig. 4.17 is embedded in the video. The message is embedded, and extraction process is similar to SVD-based approach for binary message.

Fig. 4.17 Grayscale message to be embedded

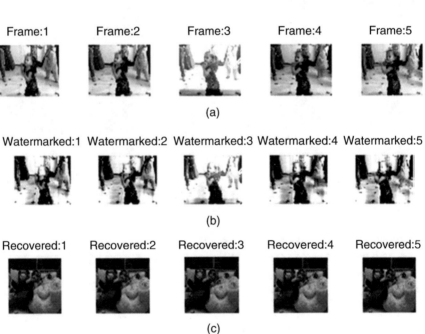

Fig. 4.18 Results of imperceptibility test of SVD-based approach for grayscale message. (**a**) Original video frames, (**b**) Watermarked video frames, (**c**) Extracted messages

4.2.1   Results of Imperceptibility Test

For imperceptibility test of SVD-based approach for grayscale message, the cover digital video with a size of 512 × 512 pixels and grayscale messages with a size of 512 × 512 pixels are taken. The one bit of message is embedded into singular values of video frame to generated watermarked frame using gain factor. Figure 4.18 shows the results of SVD-based approach for grayscale message with gain factor 100. The quality measures for SVD-based approach for grayscale message for gain factor 100 are summarized in Table 4.3.

Table 4.3 Quality measure values for SVD-based approach for grayscale message

Video frame	PSNR (dB)	Correlation
Frame 1	30.6362	0.9969
Frame 2	30.7059	0.9976
Frame 3	30.3928	0.9972
Frame 4	31.0808	0.9977
Frame 5	30.7495	0.9979

(a)

(b)

Fig. 4.19 Results of SVD-based approach for grayscale message with various gain factor values like 10, 30, 50, 70, and 90. (**a**) Watermark video Frame 1, (**b**) Extracted messages

Table 4.4 Results of SVD-based approach for grayscale message on Frame 1 using various gain factors

Gain factor (k)	PSNR (dB)	Correlation
10	53.2178	0.9985
30	40.9712	0.9984
50	35.7817	0.9983
70	32.285	0.9979
90	30.8157	0.9931
100	30.6362	0.9969

In the watermark embedding process, the gain factor is playing important role. The visual quality of watermarked video and extracted messages is depending on value of the gain factor. Figure 4.19 shows results of the method on Frame 1 considering various values of the gain factor. Table 4.4 shows the results of watermarking Frame 1 with various gain factors.

4.2.2   Results of Robustness Test

For robustness test, various types of watermarking attacks are applied on watermarked video frames, and messages are extracted from these corrupted watermarked video frames. If the message is successfully extracted from the corrupted frames,

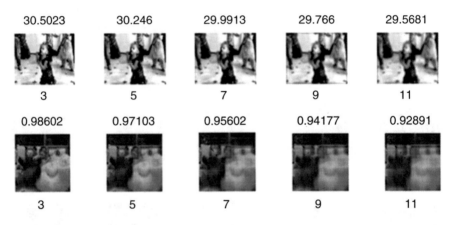

Fig. 4.20 Watermarked video frame and extracted message for SVD-based approach for grayscale message under average filtering attack with various mask sizes

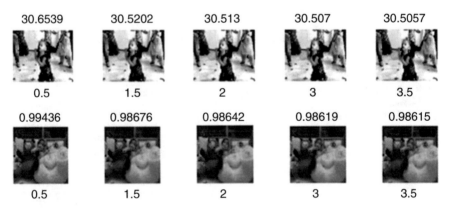

Fig. 4.21 Watermarked video frame and extracted message for SVD-based approach for grayscale message under Gaussian low-pass filtering attack with various standard deviations

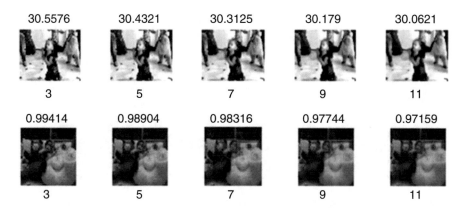

Fig. 4.22 Watermarked video frame and extracted message for SVD-based approach for grayscale message under median filtering attack with various mask sizes

Fig. 4.23 Watermarked video frame and extracted message for SVD-based approach for grayscale message under high-pass filtering attack

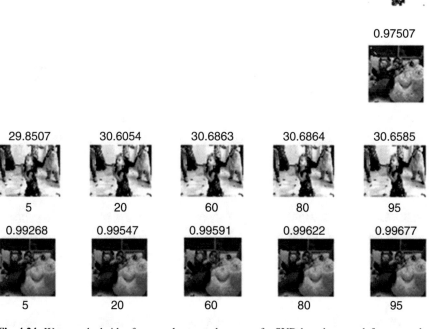

Fig. 4.24 Watermarked video frame and extracted message for SVD-based approach for grayscale message under compression attack with various quality values

Fig. 4.25 Watermarked video frame and extracted message for SVD-based approach for grayscale message under color reduction attack with various number of colors

Fig. 4.26 Watermarked video frame and extracted message for SVD-based approach for grayscale message under histogram equalization attack

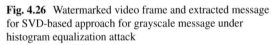

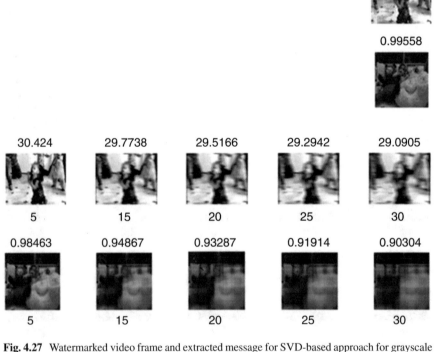

Fig. 4.27 Watermarked video frame and extracted message for SVD-based approach for grayscale message under motion blurring attack

then the approach is said to be robust. Here various types of watermarking attacks such as filtering attacks, addition of noise, geometric attacks, and other signal processing with different values are applied on watermarked video frame.

Figures 4.20, 4.21, 4.22, and 4.23 show the results of robustness test for SVD-based approach for grayscale message against different filters such as average filter, Gaussian low-pass filter median filter, and high-pass filter, respectively. Here the numbers on the upper side of the video frames show PSNR values and that in the lower side of the frames show variants of particular attacks. Similarly the numbers on the upper side of the recovered watermark show correlation values.

The compression is a very basic attack which affects any multimedia data when it is transmitted over communication channel. Figure 4.24 shows the results of robustness test for SVD-based approach for grayscale message against JPEG compression with various quality factors. Figures 4.25, 4.26, and 4.27 show the results of robustness test for SVD-based approach for grayscale message against attacks such as color reduction, histogram equalization, and motion blurring.

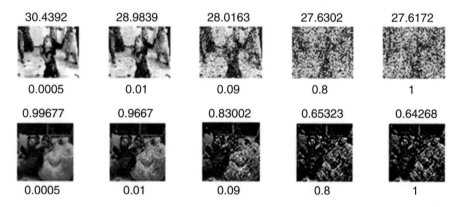

Fig. 4.28 Watermarked video frame and extracted message for SVD-based approach for grayscale message under Gaussian noise attack with zero mean and various variances

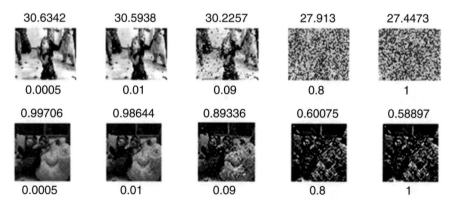

Fig. 4.29 Watermarked video frame and extracted message for SVD-based approach for grayscale message under salt and pepper noise attack with various variances

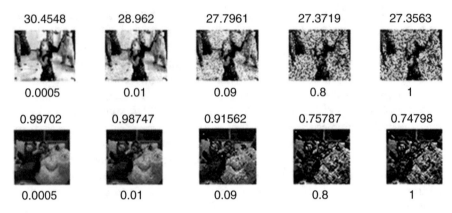

Fig. 4.30 Watermarked video frame and extracted message for SVD-based approach for grayscale message under speckle noise attack with various variances

Fig. 4.31 Watermarked video frame and extracted message for SVD-based approach for grayscale message under rotation attack with various angles

Fig. 4.32 Watermarked video frame and extracted message for SVD-based approach for grayscale message under cropping attack with various crop regions

Figures 4.28, 4.29, and 4.30 show the results of robustness test for SVD-based approach for grayscale message against various noise addition attacks such as Gaussian noise, salt and pepper noise, and speckle noise. Figures 4.31 and 4.32 show the results of robustness test for SVD-based approach for grayscale message against geometric attacks such as rotation and cropping.

4.3   Observation on Obtain Results

The following are some of the observations made after successfully implementing both embedding and extracting algorithm. Here the gain factor of 100 is assumed for the sake of observations in terms of perceptibility and robustness. The higher is the value of PSNR, the higher is the perceptibility, and the higher is the value of correlation, the higher is the robustness.

- Perceptibility in this method decreases with the increase in gain factor.
- Robustness decreases with the increase in gain factor.
- The frames look visibly fine if the resultant PSNR is above 28 dB. The message seems visibly identifiable if the resultant correlation is greater than 0.50.
- This method is fully robust against all kind of attacks in case of both binary and grayscale messages.
- This method produces more PSNR and correlation value with binary message as compared to that in the grayscale message.

4.4   Comparison with Correlation-, DCT-, and DWT-Based Approaches

The comparison of SVD-based approach with correlation-, DCT-, and DWT-based approaches is given in this section.

- At the same gain factor, perceptibility is higher than correlation-based method, and the same is less than DCT and DWT methods.
- At the same gain factor, robustness is maximum in this method.

Bibliography

1. Salwa A.K Mostafa, A. S. Tolba, F. M. Abdelkader, Hisham M. Elhindy, Video watermarking scheme based on principal component analysis and wavelet transform. IJCSNS Int. J. Comput. Sci. Netw. Secur. **9**(8), 45–52 (2009)
2. T. Al-Khatib, A. Al-Haj, L. Rajab, Video watermarking algorithms using the SVD Transform. Eur. J. Sci. Res. **30**(3), 389–401 (2009)

3. A.K. Meenal, C. Gosavi, J.P. Abhijit, Single channel watermarking for video using block based svd. Int. J. Adv. Comput. Inf. Res. **1**(2) (2012)
4. F. Huang, Z.-H. Guan, A hybrid SVD-DCT watermarking method based on LPSNR. Pattern Recogn. Lett. **25**, 1769–1775 (2004)
5. Lama Rajab, Tahani Al-Khatib, Ali Al-Haj, Hybrid DWT-SVD video watermarking. Proc. Int. Conf. Innov. Inf. Technol. 16–18 Dec 2008
6. A. Mansouri, A. Mahmoudi Aznaveh, F. Torkamani Azar, SVD-based digital image watermarking using complex wavelet transform. Sadhana **34**(3), 393–406 (2009)
7. V. Santhi, A. Thangavelu, DWT-SVD combined full band robust watermarking technique for color images in YUV color space. Int. J. Comput. Theory Eng. **1**(4), 1793–8201 (2009)

Chapter 5
Video Watermarking in Hybrid Domain

The concept of hybrid domain watermarking is presented by a researcher in the early 2000s [1–4]. Up to this point, it had been seen that individual DCT- and DWT-based technique is better in perceptibility as compared to correlation technique, while SVD-based technique is better in robustness as compared to correlation, DCT, or DWT techniques. Extra it has been observed that SVD-based technique can also be used to embed both binary and grayscale messages in the video, whereas the prior techniques were able to embed only binary messages. So in this chapter, hybridization of the two transforms, namely, DCT and DWT, and one linear algebra named SVD, is done so as to get highly perceptible watermarked video which is robust against almost all the attacks.

5.1 DCT + DWT + SVD-Based Approach for Binary Message

Here the binary message same as that used in the technique described in Chap. 4 is used for the purpose of watermarking. The hybridization of DCT, DWT, and SVD is used for the purpose of watermark embedding.

5.1.1 Watermark Embedding Process

A stepwise description of the embedding process of the proposed method is mentioned herewith.

Step 1.   Video is taken and broken into number of frames.

Step 2.   First frame is taken, and colorspace conversion is performed from RGB colorspace to the YCbCr colorspace.

© Springer International Publishing AG, part of Springer Nature 2019
A. M. Kothari et al., *Watermarking Techniques for Copyright Protection of Videos*, Signals and Communication Technology,
https://doi.org/10.1007/978-3-319-92837-1_5

Step 3.   Y component of the frame is selected for the purpose of watermarking.
Step 4.   A two-dimensional DCT is applied on the Y frame.
Step 5.   A three-level DWT is applied on the DCT transformed frame.
Step 6.   SVD is applied to both DWT transformed frame and the message frame.
Step 7.   The singular value of the frame is modified according to the singular values of the message.
Step 8.   Inverse SVD is applied to get the watermarked DWT frame.
Step 9.   Inverse three-level DWT is performed to get watermarked DCT frame.
Step 10.   Inverse DCT is applied to get the watermarked Y frame.
Step 11.   Inverse colorspace conversion is performed to get the watermarked frame.
Step 12.   Steps 2 to 11 are executed for the next frame, and the process continues until the last frame.
Step 13.   Watermarked video is obtained by combining all watermarked frames.

5.1.2   Watermark Extraction Process

The following is the stepwise representation of extracting the message at the receiver end from the video.

Step 1.   Watermarked video, which may possibly be attacked, is taken and converted into the sequence of frames.
Step 2.   First frame is taken, and colorspace conversion is performed from RGB colorspace to the YCbCr colorspace.
Step 3.   Y component of the frame is selected for the purpose of watermark extraction.
Step 4.   A two-dimensional discrete cosine transform is applied on the Y frame.
Step 5.   A three-level discrete wavelet transform (DWT) is applied on the DCT transformed frame.
Step 6.   Singular value decomposition is applied to DWT transformed frame.
Step 7.   Singular values are modified to get the watermark message back.

5.1.3   Results of Imperceptibility Test

For imperceptibility test of DCT + DWT + SVD-based approach for the binary message, the cover digital video with size of 512×512 pixels and monochrome messages with size of 512×512 pixels are taken. The one bit of message is embedded into singular values of the video frame to generate watermarked frame using gain factor. Figure 5.1 shows the results of DCT + DWT + SVD-based approach for the binary message with gain factor 100. The quality measures for DCT + DWT + SVD-based approach for the binary message for gain factor 100 are summarized in Table 5.1.

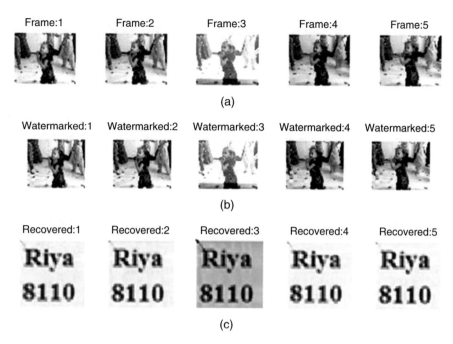

Fig. 5.1 Results of imperceptibility test of DCT + DWT + SVD-based approach for binary message. (**a**) Original video frames, (**b**) Watermarked video frames, (**c**) Extracted messages

Table 5.1 Quality measure values for DCT + DWT + SVD-based approach for binary message

Video frame	PSNR (dB)	Correlation
Frame 1	37.0173	0.9332
Frame 2	37.0311	0.9316
Frame 3	37.2750	0.8449
Frame 4	37.3222	0.9280
Frame 5	37.2390	0.9401

In the watermark embedding process, the gain factor is playing an important role. The visual quality of the watermarked video and extracted messages is depending on the value of the gain factor. Figure 5.2 shows results of the method on Frame 1 considering various values of the gain factor. Table 5.2 shows the results of watermarking Frame 1 with various gain factors.

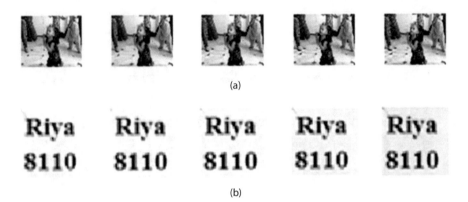

(a)

(b)

Fig. 5.2 Results of DCT + DWT + SVD-based approach for binary message with various gain factor values like 10, 30, 50, 70, and 90. (**a**) Watermark video Frame 1, (**b**) Extracted messages

Table 5.2 Results of DCT + DWT + SVD-based approach for binary message on Frame 1 using various gain factors

Gain factor (k)	PSNR (dB)	Correlation
10	56.7222	0.9516
30	45.4558	0.9522
50	41.1991	0.9518
70	38.8927	0.9508
90	37.5195	0.9487
100	37.0173	0.9332

5.1.4   Results of Robustness Test

For robustness test, various types of watermarking attacks are applied on watermarked video frames, and messages are extracted from these corrupted watermarked video frames. If the message is successfully extracted from the corrupted frames, then the approach is said to be robust. Here various types of watermarking attacks such as filtering attacks, addition of noise, geometric attacks, and other signal processing with different values are applied on watermarked video frame.

Figures 5.3, 5.4, 5.5, and 5.6 show the results of the robustness test for DCT + DWT + SVD-based approach for the binary message against different filters such as average filter, Gaussian low-pass filter median filter, and high-pass filter, respectively. Here the numbers on the upper side of the video frames show PSNR values and that in the lower side of the frames show variants of particular attacks. Similarly the numbers on the upper side of the recovered watermark show correlation values.

The compression is a very basic attack which affects any multimedia data when it is transmitted over communication channel. Figure 5.7 shows the results of the robustness test for DCT + DWT + SVD-based approach for the binary message

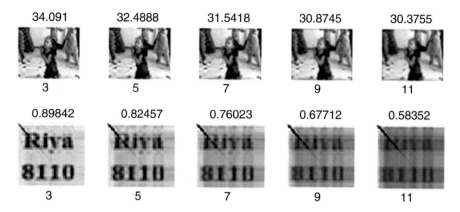

Fig. 5.3 Watermarked video frame and extracted message for DCT + DWT + SVD-based approach for binary message under average filtering attack with various mask sizes

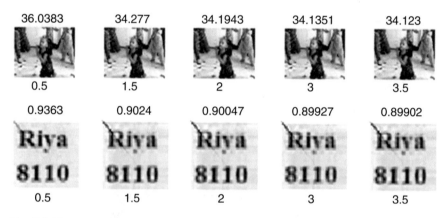

Fig. 5.4 Watermarked video frame and extracted message for DCT + DWT + SVD-based approach for binary message under Gaussian low-pass filtering attack with various standard deviations

against JPEG compression with various quality factors. Figures 5.8, 5.9, and 5.10 show the results of the robustness test for DCT + DWT + SVD-based approach for the binary message against attacks such as color reduction, histogram equalization, and motion blurring.

Figures 5.11, 5.12, and 5.13 show the results of the robustness test for DCT + DWT + SVD-based approach for the binary message against various noise addition attacks such as Gaussian noise, salt and pepper noise, and speckle noise. Figures 5.14 and 5.15 show the results of the robustness test for DCT + DWT + SVD-based approach for the binary message against geometric attacks such as rotation and cropping.

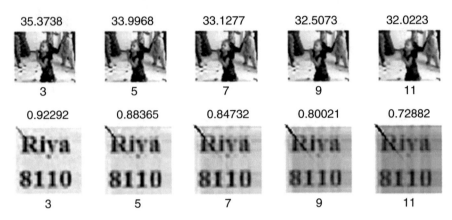

Fig. 5.5 Watermarked video frame and extracted message for DCT + DWT + SVD-based approach for binary message under median filtering attack with various mask sizes

Fig. 5.6 Watermarked video frame and extracted message for DCT + DWT + SVD-based approach for binary message under high-pass filtering attack

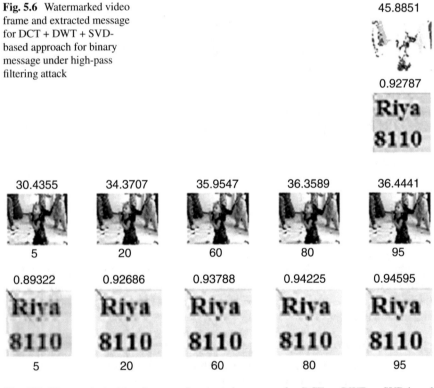

Fig. 5.7 Watermarked video frame and extracted message for DCT + DWT + SVD-based approach for binary message under compression attack with various quality values

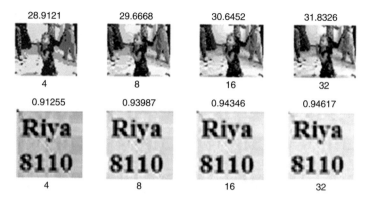

Fig. 5.8 Watermarked video frame and extracted message for DCT + DWT + SVD-based approach for binary message under color reduction attack with various numbers of colors

Fig. 5.9 Watermarked video frame and extracted message for DCT + DWT + SVD-based approach for binary message under histogram equalization attack

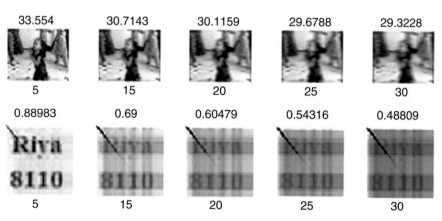

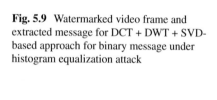

Fig. 5.10 Watermarked video frame and extracted message for DCT + DWT + SVD-based approach for binary message under motion blurring attack

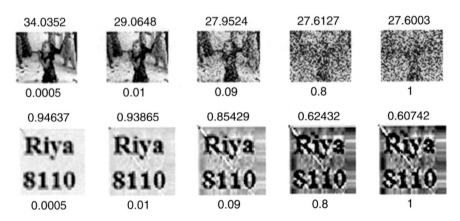

Fig. 5.11 Watermarked video frame and extracted message for DCT + DWT + SVD-based approach for binary message under Gaussian noise attack with zero mean and various variances

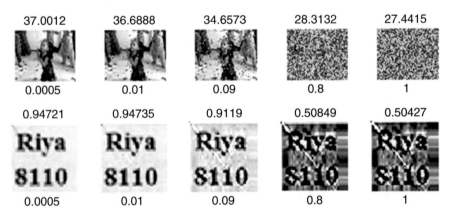

Fig. 5.12 Watermarked video frame and extracted message for DCT + DWT + SVD-based approach for binary message under salt and pepper noise attack with various variances

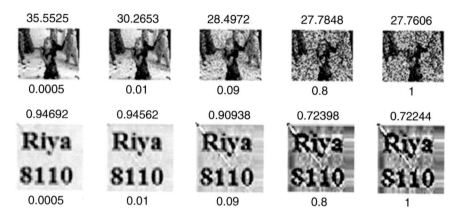

Fig. 5.13 Watermarked video frame and extracted message for SVD-based approach for binary message under spackle noise attack with various variances

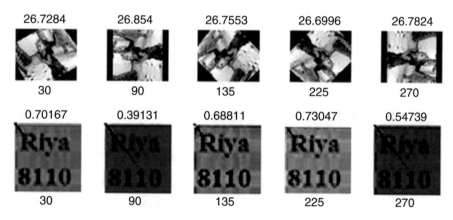

Fig. 5.14 Watermarked video frame and extracted message for DCT + DWT + SVD-based approach for binary message under rotation attack with various angles

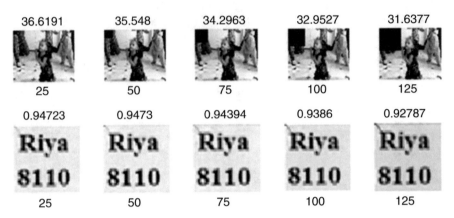

Fig. 5.15 Watermarked video frame and extracted message for DCT + DWT + SVD-based approach for binary message under cropping attack with various crop regions

5.2   DCT + DWT + SVD-Based Approach with Grayscale Message

Here the grayscale message same as that used in the SVD-based approach in Chap. 4 is used for the watermarking purpose.

5.2.1   Watermark Embedding Process

The following is the stepwise process used to embed the binary message in the video.

Step 1.   Video is taken and broken into number of frames.

Step 2.   First frame is taken, and colorspace conversion is performed from RGB colorspace to the YCbCr colorspace.

Step 3.   Y component of the frame is selected for the purpose of watermarking.

Step 4.   A two-dimensional DCT is applied on the Y frame.

Step 5.   A three-level DWT is applied on the DCT transformed frame.

Step 6.   SVD is applied to both DWT transformed frame and the message frame.

Step 7.   The singular value of the frame is modified according to the singular values of message.

Step 8.   Inverse SVD is applied to get the watermarked DWT frame.

Step 9.   Inverse three-level DWT is performed to get the watermarked DCT frame.

Step 10.   Inverse DCT is applied to get the watermarked Y frame.

Step 11.   Inverse colorspace conversion is performed to get watermarked frame.

Step 12.   Steps 2 to 11 are executed for the next frame, and the process continues until the last frame.

Step 13.   All watermarked frames are combined to get the watermarked video.

5.2.2 Watermark Extraction Process

A stepwise representation of the extraction process is explained herewith.

Step 1. Watermarked video, which may possibly be attacked, is taken and converted into the sequence of frames.

Step 2. First frame is taken, and colorspace conversion is performed from RGB colorspace to the YCbCr colorspace.

Step 3. Y component of the frame is selected for the purpose of watermark extraction.

Step 4. A two-dimensional discrete cosine transform is applied on the Y frame.

Step 5. A three-level discrete wavelet transform (DWT) is applied on the DCT transformed frame.

Step 6. Singular value decomposition is applied to DWT transformed frame.

Step 7. Singular values are modified to get the watermark message back.

5.2.3 Results of Imperceptibility Test

For the imperceptibility test of the DCT + DWT + SVD-based approach for grayscale message, the cover digital video with size of 512×512 pixels and grayscale messages with size of 512×512 pixels are taken. The one bit of message is embedded into singular values of video frame to generate watermarked frame using gain factor. Figure 5.16 shows the results of the DCT + DWT + SVD-based approach for grayscale message with gain factor 100. The quality measures for DCT + DWT + SVD-based approach for grayscale message for gain factor 100 are summarized in Table 5.3.

In the watermark embedding process, the gain factor is playing an important role. The visual quality of watermarked video and extracted messages is depending on value of the gain factor. Figure 5.17 shows results of the method on Frame 1 considering various values of the gain factor. Table 5.4 shows the results of watermarking Frame 1 with various gain factors.

5.2.4 Results of Robustness Test

For robustness test, various types of watermarking attacks are applied on watermarked video frames, and messages are extracted from these corrupted watermarked video frames. If the message is successfully extracted from the corrupted frames, then the approach is said to be robust. Here various types of watermarking attacks such as filtering attacks, addition of noise, geometric attacks, and other signal processing with different values are applied on watermarked video frame.

Fig. 5.16 Results of imperceptibility test of DCT + DWT + SVD-based approach for grayscale message. (**a**) Original video frames, (**b**) Watermarked video frames, (**c**) Extracted messages

Table 5.3 Quality measure values for DCT + DWT + SVD-based approach for grayscale message

Video frame	PSNR (dB)	Correlation
Frame 1	41.9716	0.9661
Frame 2	41.8167	0.9661
Frame 3	42.0129	0.9447
Frame 4	42.1026	0.9649
Frame 5	42.0651	0.9658

Figures 5.18, 5.19, 5.20, and 5.21 show the results of the robustness test for DCT + DWT + SVD-based approach for grayscale message against different filters such as average filter, Gaussian low-pass filter median filter, and high-pass filter, respectively. Here the numbers on the upper side of the video frames show PSNR values, and those in the lower side of the frames show variants of particular attacks. Similarly the numbers on the upper side of the recovered watermark show correlation values.

The compression is a very basic attack which affects any multimedia data when it is transmitted over communication channel. Figure 5.22 shows the results of the robustness test for DCT + DWT + SVD-based approach for grayscale message against JPEG compression with various quality factors. Figures 5.23, 5.24, and 5.25 show the results of the robustness test for DCT + DWT + SVD-based approach for

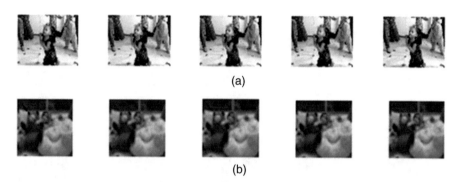

(a)

(b)

Fig. 5.17 Results of DCT + DWT + SVD-based approach for grayscale message with various gain factor values like 10, 30, 50, 70, and 90. (**a**) Watermark video Frame 1, (**b**) Extracted messages

Table 5.4 Results of DCT + DWT + SVD-based approach for grayscale message on Frame 1 using various gain factors

Gain Factor (k)	PSNR (dB)	Correlation
10	64.8356	0.9638
30	52.4004	0.9660
50	47.2649	0.9660
70	44.9679	0.9661
90	42.8519	0.9661
100	41.9716	0.9661

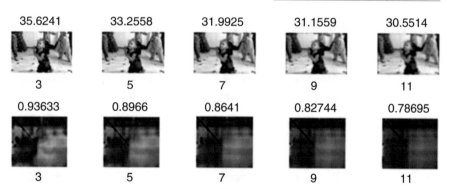

Fig. 5.18 Watermarked video frame and extracted message for DCT + DWT + SVD-based approach for grayscale message under average filtering attack with various mask sizes

grayscale message against attacks such as color reduction, histogram equalization, and motion blurring.

Figures 5.26, 5.27, and 5.28 show the results of the robustness test for DCT + DWT + SVD-based approach for grayscale message against various noise addition attacks such as Gaussian noise, salt and pepper noise, and speckle noise. Figures 5.29 and 5.30 show the results of the robustness test for DCT + DWT + SVD-based approach for grayscale message against geometric attacks such as rotation and cropping.

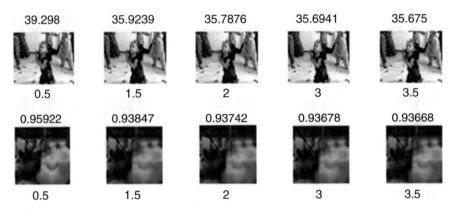

Fig. 5.19 Watermarked video frame and extracted message for DCT + DWT + SVD-based approach for grayscale message under Gaussian low-pass filtering attack with various standard deviations

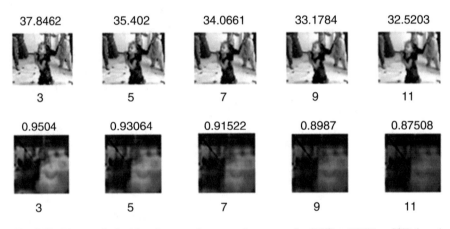

Fig. 5.20 Watermarked video frame and extracted message for DCT + DWT + SVD-based approach for grayscale message under median filtering attack with various mask sizes

Fig. 5.21 Watermarked video frame and extracted message for DCT + DWT + SVD-based approach for grayscale message under high-pass filtering attack

85.4026

0.85928

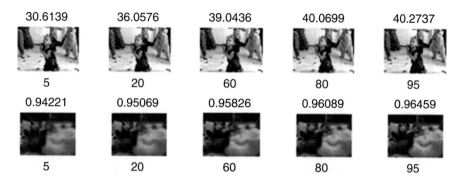

30.6139	36.0576	39.0436	40.0699	40.2737
5	20	60	80	95
0.94221	0.95069	0.95826	0.96089	0.96459
5	20	60	80	95

Fig. 5.22 Watermarked video frame and extracted message for DCT + DWT + SVD-based approach for grayscale message under compression attack with various quality values

28.9658	29.8629	31.0314	32.4671
4	8	16	32
0.94431	0.95731	0.96506	0.96617
4	8	16	32

Fig. 5.23 Watermarked video frame and extracted message for DCT + DWT + SVD-based approach for grayscale message under color reduction attack with various numbers of colors

Fig. 5.24 Watermarked video frame and extracted message for DCT + DWT + SVD-based approach for grayscale message under histogram equalization attack

28.9217

0.96429

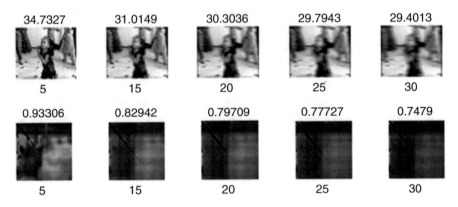

Fig. 5.25 Watermarked video frame and extracted message for DCT + DWT + SVD-based approach for grayscale message under motion blurring attack

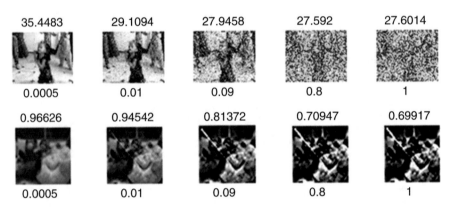

Fig. 5.26 Watermarked video frame and extracted message for DCT + DWT + SVD-based approach for grayscale message under Gaussian noise attack with zero mean and various variances

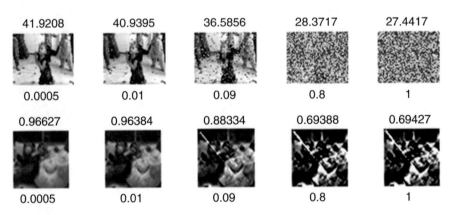

Fig. 5.27 Watermarked video frame and extracted message for DCT + DWT + SVD-based approach for grayscale message under salt and pepper noise attack with various variances

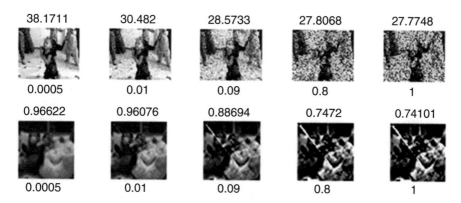

Fig. 5.28 Watermarked video frame and extracted message for DCT + DWT + SVD-based approach for grayscale message under speckle noise attack with various variances

Fig. 5.29 Watermarked video frame and extracted message for DCT + DWT + SVD-based approach for grayscale message under rotation attack with various angles

Fig. 5.30 Watermarked video frame and extracted message for DCT + DWT + SVD-based approach for grayscale message under cropping attack with various crop regions

5.3   Observation on Obtain Results

The following are some of the observations made after successfully implementing both embedding and extracting algorithm. Here the gain factor of 100 is assumed for the sake of observations in terms of perceptibility and robustness. The higher is the value of PSNR, the higher is the perceptibility, and the higher is the value of correlation, the higher is the robustness.

- Perceptibility in this method decreases with the increase in gain factor.
- Robustness decreases with the increase in gain factor.
- The frames look visibly fine if the resultant PSNR is above 28 dB. The message seems visibly identifiable if the resultant correlation is greater than 0.50.
- This method is fully robust against all kind of attacks.

5.3.1   Comparison with DCT-, DWT-, and SVD-Based Approaches

The comparison of DCT + DWT + SVD-based approach with DCT-, DWT-, and SVD-based approaches is given in this section.

- Perceptibility in this method is considerably highest among all methods at the same gain factor.
- Robustness achieved is considerably higher than correlation, DCT-, and DWT-based methods and slightly less than SVD-based method.

Bibliography

1. R.B. Wolfgang, C.I. Podilchuk, Perceptual watermarks for digital images and video. Proc. IEEE **87**(7) (1999)
2. G. C. Langelaar, I. Setyawan, R. L. Lagendijk, Watermarking of digital image and video data – A state of art review. IEEE Signal Proc. Mag. 20–46 Sept 2000
3. C.I. Podilchuk, E.J. Delp, Digital watermarking: Algorithms and applications. IEEE Signal Proc. Mag. **18**(4), 33–46 (2001)
4. R. Chandramouli, N.D. Memon, M. Rabbani, Digital watermarking, in *Encyclopedia of Imaging Science and Technology*, (Wiley, New York, 2002)

Chapter 6
Video Watermarking in Sparse Domain

Up to this point, it had been seen that individual DCT, DWT, or hybridization of various transforms DCT + DWT + SVD is used for copyright protection of videos. In these approaches, frequency coefficients of cover video frame are modified by watermark information to get watermarked video frame.

Around 2007, researchers are introduced new watermarking approach where watermark information is inserted into sparse data of cover medium [1]. Thereafter, many approaches in sparse domain are presented by various researchers [2–7]. The main part of this sparse domain watermarking scheme is used of compressive sensing (CS) theory. The CS theory is described by Donoho and Candes in 2005, based on the sparsity property of the image [8, 9]. This theory converts image into its sparse measurements using image transforms and random measurement matrix. When any image transform is applied on the image, then the image is converted into its sparse coefficients. These coefficients are multiplied with measurement matrix (which is random in nature) to get sparse measurements of the image.

The reconstruction of the image from its sparse measurements can be done using various optimization techniques. Researchers are described with various techniques based on L_1 minimization and greedy approach for reconstruction of the image in the literature [10–13]. The CS theroy converts image in its encrypted form in term of sparse data. Thus, it improves security of image. Due to this property of CS theory, it is used in watermarking.

6.1 CS Theory-Based Video Watermarking Approach

In sparse domain watermarking, watermark image is inserted into sparse measurements of the cover image, and then image reconstruction is applied on watermarked sparse measurements to get a watermarked image. The generation of sparse measurements of the cover image and reconstruction of watermarked image from watermarked sparse measurements is done using CS theory process.

© Springer International Publishing AG, part of Springer Nature 2019
A. M. Kothari et al., *Watermarking Techniques for Copyright Protection of Videos*, Signals and Communication Technology,
https://doi.org/10.1007/978-3-319-92837-1_6

6.1.1   Watermark Embedding Process

A stepwise description of the embedding process of the proposed method is mentioned herewith.

Step 1. Video is taken and broken into number of frames.

Step 2. First frame is taken, and colorspace conversion is performed from RGB colorspace to the YCbCr colorspace.

Step 3. Y component of the frame is selected for the purpose of watermarking.

Step 4. Obtain sparse coefficients Cx of video frame by multiplying video frame V with wavelet basis matrix and its inverse as follows:

$$Vx = \Psi \times V \times \Psi' \qquad (6.1)$$

where Vx denotes sparse coefficients of the host image and Ψ is the wavelet basis matrix.

Step 5. Generate measurement matrix A using standard normal Gaussian distribution with mean $= 0$ and variance $= 1$.

Step 6. Multiply the measurement matrix A with sparse coefficients Cx to get sparse measurements Vy of video frame. Break these sparse measurements into non-overlapping blocks of size 8×8.

Step 7. Generate two highly uncorrelated PN sequences using noise generator; each of the size equals to the size of the block.

Step 8. Create the watermark mask WM based on PN sequences, watermark bits, and size of sparse measurements using the following conditions:

(a) If watermark bit is zero, add PN sequence zero at that block position of the mask.

(b) Otherwise, generate mask using PN sequence one.

(c) Repeat this process for all blocks of medical image.

Step 9. Insert watermark mask into sparse measurements using gain factor k to get watermarked sparse measurements of the video frame.

$$WVy = Vx + k \times WM \qquad (6.2)$$

Where, WVy is the watermarked sparse measurements of video frame, and WM is a watermark mask.

Step 10. Apply the measurement matrix A with the help of OMP algorithm to watermarked sparse measurements to get the watermarked sparse coefficients WVx of the video frame.

$$WVx = OMP(WVy,A) \qquad (6.3)$$

Where, WVx is watermarked sparse coefficients of video frame.

Step 11. Finally, apply inverse wavelet basis matrix with its original form on the watermarked sparse coefficients to get the watermarked video frame V^*.

$$V^* = \Psi' \times WVx \times \Psi \qquad (6.4)$$

Where, $V*$ is the watermarked video frame.

Step 12. Inverse colorspace conversion is performed to get watermarked frame.

Step 13. Steps 2 to 121 are executed for the next frame, and the process continues until the last frame.

Step 14. Watermarked video is obtained by combining all the watermarked frames.

6.1.2 Watermark Extraction Process

The following is the stepwise representation of extracting the message at the receiver end from the video.

Step 1. The watermarked video, which may possibly be attacked, is taken and converted into the sequence of frames.

Step 2. The first frame is taken, and colorspace conversion is performed from the RGB colorspace to the YCbCr colorspace.

Step 3. The Y component of the frame is selected for the purpose of watermark extraction.

Step 4. Apply Step numbers 4–6 of the watermark embedding process on watermarked video frame to get the sparse measurements $Vy*$. Break these sparse measurements into non-overlapping blocks of the size of 8×8.

Step 5. Generate the same two highly uncorrelated PN sequences which are generated during watermark embedding process.

Step 6. Extract the watermark bits from sparse measurements of the watermarked video frame based on the following conditions:

$$Sequence\_1 = corr2\left(Vy^*, PN\_Sequence\_1\right) \qquad (6.5)$$

$$Sequence\_2 = corr2\left(Vy^*, PN\_Sequence\_0\right) \qquad (6.6)$$

Step 7. If Sequence_1 < Sequence_2, then set watermark bit as 0. Otherwise, set watermark bit as bit 1.

Step 8. Reshape the extracted bit vector to obtain extracted watermark message.

6.1.3 Results of Imperceptibility Test

For imperceptibility test of CS theory-based approach, the cover digital video with the size of 256 × 256 pixels and monochrome messages with the size of 32 × 32 pixels are taken. The one bit of message is embedded into an 8 × 8 block of sparse

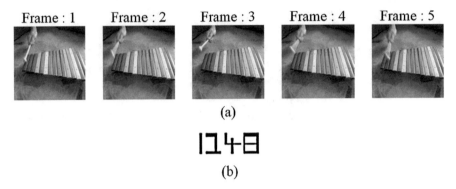

(b)

Fig. 6.1 (**a**) Test cover video frames, (**b**) Watermark message

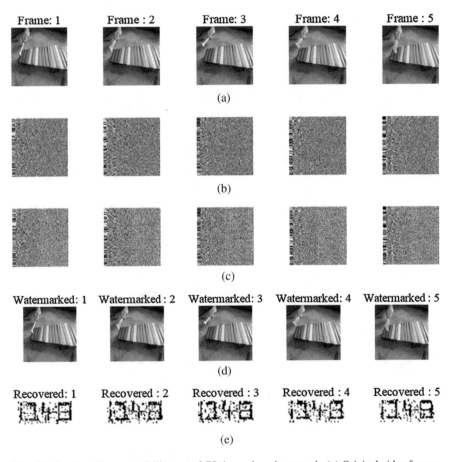

Fig. 6.2 Results of imperceptibility test of CS theory-based approach. (**a**) Original video frames, (**b**) Sparse measurements of video frames, (**c**) Watermarked sparse measurements of video frames, (**d**) Watermarked video frames, (**e**) Extracted messages

Table 6.1 Quality measure values for CS theory-based approach

Video frame	PSNR (dB)	Correlation
Frame 1	32.0115	0.8706
Frame 2	32.0704	0.8437
Frame 3	31.9724	0.8625
Frame 4	32.0792	0.8747
Frame 5	32.1830	0.8881

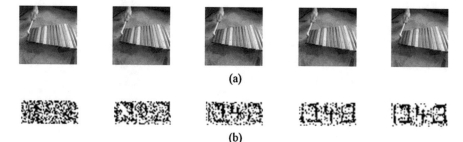

Fig. 6.3 Results of CS theory-based approach with various gain factor values like 10, 30, 50, 70, and 90. (**a**) Watermark video Frame 1, (**b**) Extracted messages

Table 6.2 Results of CS theory-based approach on Frame 1 using various gain factors

Gain factor (k)	PSNR (dB)	Correlation
10	38.5295	0.5593
30	37.8105	0.7022
50	36.3728	0.7493
70	34.6571	0.8073
90	34.4362	0.8383
100	32.0115	0.8706

measurements of video frame to generate a watermarked frame using a gain factor. Figure 6.1 shows test cover video frames and watermark message. Here five frames of the video are used for better representation of results (Fig. 6.2).

Figure 6.1 shows the results of CS theory-based approach with gain factor 100. The quality measures for CS theory-based approach for gain factor 100 are summarized in Table 6.1.

In the watermark embedding process, the gain factor is playing an important role. The visual quality of watermarked video and extracted messages is depending on the value of gain factor. Figure 6.3 shows results of the method on Frame 1 considering various values of the gain factor. Table 6.2 shows the results of watermarking Frame 1 with various gain factors.

Fig. 6.4 Results of the
robustness test for CS
theory-based watermarking
approach against
watermarking attacks. (**a**)
Compression attack
($Q = 5$), (**b**) Compression
attack ($Q = 20$), (**c**)
Compression attack
($Q = 60$), (**d**) Compression
attack ($Q = 80$), (**e**)
Compression attack
($Q = 95$), (**f**) Gaussian
noise attack, (**g**) Salt and
pepper noise attack, (**h**)
Speckle noise attack, (**i**)
Median filter attack, (**j**)
Mean filter attack, (**k**)
Histogram equalization
attack, (**l**) Rotation attack
($Q = 95$), (**m**) Cropping
attack, (**n**) Motion blurring
attack

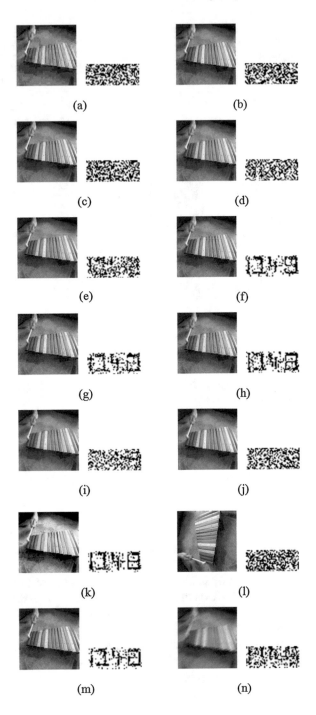

Watermarking Attacks	PNSR (dB)	NC
Compression Attack ($Q = 5$)	23.93	0.5216
Compression Attack ($Q = 20$)	29.72	0.5229
Compression Attack ($Q = 60$)	33.18	0.5485
Compression Attack ($Q = 80$)	34.96	0.5903
Compression Attack ($Q = 95$)	36.75	0.6792
Gaussian Noise (mean = 0, variance = 0.001)	27.99	0.8531
Salt & Pepper Noise (variance = 0.005)	26.50	0.8167
Speckle Noise (variance = 0.005)	27.60	0.8194
Median Filter (mask size = 3×3)	33.22	0.5997
Mean Filter (mask size = 3×3)	30.32	0.5296
Histogram Equalization	18.58	0.8571
Rotation	11.05	0.4663
Cropping	25.49	0.8679
Motion Blurring	21.46	0.6887

Table 6.3 PSNR and NC values for Robustness Results for CS theory based Watermarking Approach against Watermarking Attacks

6.1.4 Results of Robustness Test

For robustness test, various types of watermarking attacks are applied on watermarked video frames, and messages are extracted from these corrupted watermarked video frames. If the message is successfully extracted from the corrupted frames, then the approach is said to be robust. Here various types of watermarking attacks such as filtering attacks, addition of noise, geometric attacks, and other signal processing with different values are applied on watermarked video frame. Figure 6.4 shows the results of robustness test for CS theory-based approach for binary message against different watermarking attacks. Here the results of watermarking attacks applied on Frame 1 are demonstrated. The corresponding PSNR and correlation values are summarized in Table 6.3.

6.1.5 Observation on Obtain Results

The following are some of the observations made after successfully implementing both embedding and extracting algorithm. Here the gain factor of 100 is assumed for the sake of observations in terms of perceptibility and robustness. The higher is the value of PSNR, the higher is the perceptibility, and the higher is the value of correlation, the higher is the robustness.

- Perceptibility in this method decreases with the increase in gain factor.
- Robustness decreases with the increase in gain factor.
- This method is fully robust against Gaussian noise attack, speckle noise attack, cropping attack, histogram equalization attack, and salt and pepper noise attack.
- This method is partially robust against motion blurring attack. This method is not robust against compression attack, median filter attack, mean filter attack, and rotation attack.

Bibliography

1. M. Sheikh and R. Baraniuk, Blind error free detection of transform domain watermarks. IEEE Int. Conf. Image. Proc. 5 Sept 2007
2. F. Tiesheng, L. Guiqiang, D. Chunyi, W. Danhua, A digital image watermarking method based on the theory of compressed sensing. Int. J. Autom. Control Eng. 2(2), 56–61 (2013)
3. A. Sreedhanya, K. Soman, Ensuring security to the compressed sensing data using a Steganographic approach. Bonfring Int. J. Adv. Image Proces. 3(1), 1–7 (2013)
4. M. Fakhr, Robust watermarking using compressed sensing framework with application to MP3 audio. Int. J. Multimed. Appl. (IJMA) 4(6), 27–43 (2012)
5. M. Raval, M. Joshi, P. Rege and S. Parulkar, Image tampering detection using compressive sensing based watermarking scheme. Proc. MVIP 2011. 2011
6. X. Zhang, Z. Qian, Y. Ren, G. Feng, Watermarking with flexible self-recovery quality based on compressive sensing and compositive reconstruction. IEEE Trans. Inf. Forensic. Secur. 6(4), 1123–1232 (2011)
7. M. Tagliasacchi, G. Valenzise, S. Tubaro, G. Cancelli, M. Barni, A compressive sensing based watermarking scheme for sparse image tampering identification. Proc. ICIP 2009. 1265–1268 (2009)
8. D. Donoho, Compressed sensing. IEEE Trans. Inf. Theory 52(4), 1289–1306 (2006)
9. E. Candes, Compressive Sampling. Proc. Int. Congr. Math. 1–20 (2006)
10. R. Baraniuk, Compressive Sensing. IEEE Signal Process. Mag. 24, 118–124 (2007)
11. E. Candes and J. Romberg, L1-Magic: recovery of sparse signals via convex programming. 1–19 (2005)
12. J. Tropp, A. Gilbert, Signal recovery from random measurements via orthogonal matching pursuit. IEEE Trans. Inf. Theory 53(12), 4655–4666 (2007)
13. A. Gilbert, M. Strauss, J. Tropp, R. Vershynin, One sketch for all: Fast algorithms for compressed sensing. 39th ACM Symp. Theory Comput. (STOC). 237–246 (2007)

Chapter 7
Conclusion and Future Direction

This book provides various watermarking approaches for copyright protection of digital videos. The book provides watermarking approaches in various processing domains such as spatial, transform, hybrid, and sparse. The results are demonstrated for color video frames using monochrome watermark message and grayscale watermark message. The results also indicated that hybrid domain watermarking approach has provided much better results in terms of imperceptibility and robustness compared to results of other domain approaches.

The computational time is one of the important parameters for watermarking when it is designed for copyright protection. The computational time is the total time required for watermark embedding process and watermark extraction process. The computational time of presented approaches in the book is summarized in Fig. 7.1. Figure 7.1 shows that hybrid domain watermarking approaches are required less computational time compared to other approaches. This indicated that for copyright protection of videos, hybrid watermarking approaches are the best solution.

There is a lot of research still available in the area of digital video watermarking. The development in various advanced image transforms such as fast discrete curvelet transform, finite ridgelet transform, and discrete contourlet transform overcome some limitation of conventional image transforms like DCT, DFT, DWT, etc. thus, design of video watermarking algorithm using these transforms is one of the open and prominent suggested research area which can also be used for intelligent technique such as bio inspired algorithms, machine learning algorithms, and deep learning algorithms in video watermarking approaches to improve the results of traditional watermarking approaches.

© Springer International Publishing AG, part of Springer Nature 2019
A. M. Kothari et al., *Watermarking Techniques for Copyright Protection of Videos*, Signals and Communication Technology,
https://doi.org/10.1007/978-3-319-92837-1_7

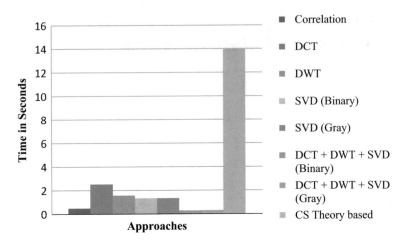

Fig. 7.1 Computational time of video watermarking approaches

Index

Printed in the United States
By Bookmasters